HPC@Green IT

Ralf Gruber · Vincent Keller

HPC@Green IT

Green High Performance Computing Methods

Foreword by Erich Strohmaier

 Springer

Dr. Ralf Gruber
Ecole Polytechnique Fédérale de
Lausanne (EPFL)
Laboratoire Mécanique
1015 Lausanne
Switzerland
ralf.gruber@epfl.ch

Dr. Vincent Keller
Fraunhofer-Gesellschaft
Inst. Algorithmen und
Wissenschaftliches Rechnen SCAI
Schloss Birlinghoven
53754 Sankt Augustin
Germany
Vincent.Keller@a3.epfl.ch

ISBN 978-3-642-01788-9   e-ISBN 978-3-642-01789-6
DOI 10.1007/978-3-642-01789-6
Springer Heidelberg Dordrecht London New York

Library of Congress Control Number: 2009939538

Cover design: KuenkelLopka GmbH, Heidelberg

Printed on acid-free paper

Springer is part of Springer Science+Business Media (www.springer.com)

Foreword

Making the most efficient use of computer systems has rapidly become a leading topic of interest for the computer industry and its customers alike. However, the focus of these discussions is often on single, isolated, and specific architectural and technological improvements for power reduction and conservation, while ignoring the fact that power efficiency as a ratio of performance to power consumption is equally influenced by performance improvements and architectural power reduction. Furthermore, efficiency can be influenced on all levels of today's system hierarchies from single cores all the way to distributed Grid environments. To improve execution and power efficiency requires progress in such diverse fields as program optimization, optimization of program scheduling, and power reduction of idling system components for all levels of the system hierarchy.

Improving computer system efficiency requires improving system performance and reducing system power consumption. To research and reach reasonable conclusions about system performance we need to not only understand the architectures of our computer systems and the available array of code transformations for performance optimizations, but we also need to be able to express this understanding in performance models good enough to guide decisions about code optimizations for specific systems. This understanding is necessary on all levels of the system hierarchy from single cores to nodes to full high performance computing (HPC) systems, and eventually to Grid environments with multiple systems and resources. A systematic, coherent, and measurement-based approach is necessary to cover the large system and application space we face nowadays.

Code optimization strategies are quite different for the various levels of system architectures. Single-core optimizations have long been researched, but due to the memory wall and complex core architectures still remain of crucial importance. With the rapid shift to multi-core technology the degree of concurrency in our systems and codes is rapidly raising. This puts a strong emphasis on node and system-wide parallel efficiency even for application domains, which in the area of single core processors were able to ignore parallel execution. On the machine room and enterprise level, Grid environments are now routinely used to optimize resource utilization. Here not only code optimizations, but optimal resource selection and parallel job scheduling are necessary.

More and more system components allow dynamically selecting a trade-off between power consumption and performance. On the smallest scale, frequencies of computer cores or processors are often user-tunable while at larger scales complete nodes of a system can rapidly be taken in and out of power-saving, idle states. To take full advantage of these capabilities, we need performance models which allow program specific decisions such as which and how many resources to use and how to select optimal frequencies and power consumption for individual components.

The current book HPC@Green IT by Ralf Gruber and Vincent Keller is unique in addressing all of these topics in a coherent and systematic way and in doing so fills an important gap. The methods presented and their integration have the potential to influence power efficiency and consumption on various scales of the system architectures and should ultimately help the greening of HPC.

Dr. Erich Strohmaier
Lawrence Berkeley National Laboratory, August 2009

Preface

Advanced economical and ecological information technology methods are presented with practical examples. A special effort is made to show how the energy consumption per job can be reduced: Adjusting processor frequency to application needs, optimizing applications on cores, nodes, clusters and Grids, choosing the best suited resource to execute a job, stopping inefficient resources, and adapting new machines to the application community. For this purpose, the applications and the computational and communication resources are parameterized, running jobs and the status of the resources are monitored, and the collected data is stored in a data base. At run time the input queues of all the eligible resources are accessed by the metascheduler and used to predict job waiting time. Models are applied to predict CPU time and network needs. A cost function is then defined to estimate the real cost of the application on the resources, including costs related to CPU time, licensing, energy consumption, data transfer, and waiting costs. The lowest cost resource is chosen. If already reserved, the application is submitted to the next best one. The environment can be used by computing centers to optimally schedule a Grid of computers.

The historical data is used to detect inefficient resources that should be decommissioned, and to predict new resources that would be best suited to the application community. The complexity analysis needed to estimate the CPU time can help to predict an improper behavior of an application, and can give hints to improve it.

All this information is used to reduce energy consumption directly or indirectly. Thus, it is possible to automatically adjust the frequency of each node to the needs of the application task, reducing processor energy consumption. Another energy consumption reduction can be achieved by increasing the efficiencies of the nodes and of the internode communication networks, done by an optimal scheduling of applications to well suited resources in a computational Grid. It is also possible to detect energy inefficient resources that should be stopped, and replaced by those adjusted to the needs of the Grid user community. And last but not least, optimizing the applications on cores, nodes, clusters, and on a Grid reduces their run time, and lowers energy consumption per run. Often, this latter energy reduction step is the most profitable.

All those actions are agnostic to the applications, thus, there is no need to modify a program.

Lausanne, Switzerland   *Ralf Gruber*
June 2009   *Vincent Keller*

Acknowledgements

First and foremost, we would like to thank Marie-Christine Sawley and Michel Deville. The realization of this book would not have been possible without their continuous help. Special thanks go to Trach-Minh Tran and Michael Hodous who contributed in various chapters.

A book about code optimization, prediction models, energy saving through application-oriented computing – or simply computational science – is a meeting place. We must emphasize that computational science requires skills and knowledge from fields as different as physics, mathematics, chemistry, mechanics, or computer science. We present in this book applications from plasma physics and computational fluid dynamics. Without the help of their developers and users, it would not have been possible to finish those parts of the book that validate the theories, i.e. the examples. We would like to thank (in no specific order), Wilfred Anthony Cooper for his outstanding help in the TERPSICHORE, VMEC and LEMan codes, Martin Jucker on TERPSICHORE, Nicolas Mellet on LEMan, Roland Bouffanais, Christoph Bosshard, Sohrab Kethari and Jonas Lätt on the SpecuLOOS code, and Emmanuel Leriche and Marc-Antoine Habisreutinger on the Helmholtz 3D solver. Thanks also go to Silvio Merazzi for his continuous availability when using the Baspl++ software. Michael Hodous is the author of the Section 5.2.8 in Chapter 5 on code porting, and we also acknowledge him for his contributions to the history of supercomputing, and for rephrasing our unidiomatic English. Francis Lapique contributed with chapter 6.1.3 on the use of graphics processors as an HPC machine.

Monitoring an application is not always straightforward. It is mandatory to be familiar with the target architecture, and to benchmark, and fine low-level monitor it. Trach-Minh Tran delivered most of the benchmark results of the kernels. Thanks go to Guy Courbebaisse for his benchmarks on the Intel Atom architecture, and to Basile Schaeli for his MPI library version that allows a user to monitor the MPI traffic of an application.

All the applications presented in this book have run on the Pleiades cluster at the Institute of Mechanical Engineering at EPFL, managed by Ali Tolou; many thanks to him. We would like to thank the head of the central computing center at EPFL (DIT) Jean-Claude Berney and Christian Clémençon to let us run SpecoLOOS on the full Blue Gene/L machine.

Parts of the ïanos project presented in Chapter 8 would not be possible without the collaboration with European institutions and research scientists through the CoreGRID network of excellence. Thanks go to Wolfgang Ziegler, Philipp Wieder, Hassan Rasheed and Oliver Wäldrich.

Thanks go to several institutions for their support: the École Polytechnique Fédérale de Lausanne, and, specifically, the Laboratory of Computational Engineering (LIN-EPFL), the Centre de Recherche en Physique des Plasmas (CRPP-EPFL), and the central computing center (DIT-EPFL), the Swiss National Supercomputing Centre (CSCS-ETHZ) in Manno, the University of Bonn (B-IT) and the Fraunhofer Institute SCAI in Sankt-Augustin, Germany.

We thank all the reviewers for their perfect remarks and corrections. The book has strongly improved thanks to their interventions.

Last but not least, we would like to thank the two most important persons: Heidi and Céline to have supported us for so long. Both on your side, you changed our lives for the best.

Contents

Chapter 1
Introduction

*"Because we don't think about future generations, they will
never forget us."*
Henrik Tikkanen, Swedish writer (1924 –1984)

Abstract In 2006, Joseph A. Stanislaw (American economist) wrote "Energy: The
21st Century Challenge". Based on the same observation (one of the main concerns of the inhabitants of the Earth), this first chapter introduces the context of the
book.

1.1 Basic goals of the book

When performing computer simulation for weather forecast, airplane design, new
material search, or experimental data interpretation, the right choice of the physics
model, its mathematical description, and the applied numerical methods used are
crucial to get as close as possible to experimental measurements and observations.
These topics are not treated in the book. We concentrate on application implementation aspects related to an optimal usage of the underlying computer architectures and
their resources. One of the major concerns is the energy consumption that should
be reduced in future. Ideas such as adapting the computer to application needs,
submission of the application to the most suitable machine, or simply by optimizing
the application implementation can lead to substantial energy savings, and to faster
results.

1.2 What do I get for one Watt today ?

Despite technology evolution towards more and more integrated circuitries, the
energy consumption of personal computers (PC) has stagnated for more than 10
years at about 200 Watts per node, and laptops are at 50 Watts. Today, we talk
about 45 nm technology; soon 32 nm will appear on the market. Each step towards
denser VLSI further reduces the energy consumption of a given circuit. In 2000, a
RISC node had just one processor and one core. Ten years later, there are nodes
with 4 processors each having 6 cores, but the cycle period of around 3 GHz

R. Gruber, V. Keller, *HPC@Green IT*,
DOI 10.1007/978-3-642-01789-6_1, © Springer-Verlag Berlin Heidelberg 2010

remained constant. The energy saving due to new technology is largely offset by the many-core architecture.

The 200 Watts consumed in a computer are roughly equally due to processors, main memory subsystems, and the rest (I/O, networks, graphic card, etc). Most modern processors have the possibility to change core frequency [4, 81] stepwise. The processors go to sleep at lowest frequency when it is not used during a certain time. This reduces drastically energy consumption since it is proportional to the frequency and to the voltage squared [107]. Today's processors reduce frequency when there is no action, but if the operating system (OS) detects an activity (mainly based on the CPU load), the frequency governor orders the processor to run at full speed. For instance, when preparing a letter or an email, or by surfing on the Web, the cores could run at low frequency, or some of them could even be stopped. The same is true if one executes high performance computing applications that are dominated by main memory access, or by I/O operations that can, for instance, occur in database-oriented applications. It is possible to run each core in a processor at top frequency, at sleeping frequency, or at a frequency in between. This can be used to adapt it to the needs of an application during execution. Sometimes, it would even be good to stop one or more cores in a node.

The information technology (IT) energy saving potential is high. Presently, IT takes more than 5% of the worldwide produced electricity, tendency is up. Controlling all parts in a computer would lead to a substantial economical and ecological advantage. The first step, i.e. adapting the processor frequencies, can already lead to a power reduction by up to 30%. In the near future the parts besides the cores should also be made controllable to save energy even more.

Let us compare the price/performance and performance/Watt ratios of an old Cray-2 with a node of today. The four processor Cray-2 came out in 1985 with a cycle period of 4.1 ns, and 2 GB of common main memory. Each processor was able to produce an add and a multiply operation in one cycle period, leading to a theoretical peak performance of close to 2 GF/s. Its cost was 40 millions of US dollars, and over 100 M$ if inflation is counted. The Cray-2 was liquid cooled and consumed 200 kW. Supposing an electricity price of 0.1 $ per kWh, the yearly electricity bill was about half a percent of the investment costs. Typically, applications could run at half of peak performance. Today's nodes typically cost 2000 $ and consume 200 W. With again 0.1 $ per kWh, the yearly electricity costs is 10 % of investment, with cooling about 15%. Point-to-point applications typically run at around 2% of peak, that is a 25 times smaller efficiency compared to a Cray-2. Nodes today have peak performances of around 40 GF/s (1GF/s= 10^9 operations per second), 1 GF/s real performance for main memory access dominated codes, thus, about the same performance as on a Cray-2. When applying Moore's law (factor 2^{18} in 27 years) to the cost and the energy consumption, a node with a Cray-2 performance today should cost 400 $ (off by a factor of 4 when inflation is accounted) and energy consumption should be at 1 Watt (off by a factor of 100).

1.3 Main memory bottleneck

Another important part of a computer is main memory. From 2000 to 2010 the memory size went from 0.5 GB (1 GB $= 10^9$ Bytes) to 16 GB per node. The main memory bandwidth went up, but not as fast as the size. In many applications main memory access is the bottleneck. Iterative matrix solvers for instance are based on sparse matrix times vector (BLAS2) operations demanding high main memory bandwidths of around one operand per operation. Vector machines such as the Cray X1E and the NEC SX, constructed for data intensive high performance computing (HPC) applications, can deliver such bandwidths. To do so, the memory is cut into banks that can be accessed in parallel. These complex main memory subsystems lead to high costs.

In cache based RISC architectures, data transfer from main memory to the highest level cache is done with less than one Byte per operation, thus delivering a tenth or less of vector machine memory bandwidth efficiencies, not enough for BLAS2 operations. If data is in cache the access of data is very fast (7 cycle periods in a Pentium 4) and the bandwidth reaches one operand per cycle period as in vector machines. This makes it clear that cache misses should be kept as small as possible, or, data in cache should be used as often as possible. This is one of the major goals when optimizing programs on a core.

1.4 Optimize resource usage

Optimize usage of resources implies a very thorough monitoring of the application during execution. For instance, in the ïanos [38, 88, 64, 39, 44, 110] environment, information on the behavior of the application is stored in a data base during execution, and reused later to predict on which computer a submitted application should best be executed. This decision is taken after having estimated the "costs" of the application on the available resources. This objective function includes costs related to CPU time, licensing of application software from individual software vendors (ISV), energy consumption, data transfer, and waiting costs. Parameters can be triggered to optimize with respect to one or several of those costs. A computing centre for instance would like to best use their installed resources by producing as many productive cycles as possible, whereas users would like to get results as rapidly as possible. Weather forecasts must be presented at the evening television news, or shipped to the air traffic control, needing a dedicated usage of the resources in a given time window. Other users would like to run their applications at lowest costs, or would like to run on nodes consuming least energy.

The monitoring information captured by ïanos can be reused to predict if an older computer or cluster should be decommissioned. In one case, we found that the energy cost to run applications on an older cluster was larger than the overall cost on a new cluster. In the meantime, this old machine has been replaced by a new one that is now well suited for run the applications of its user community.

The ïanos environment can also be used to detect improper behaviors of programs on one node, or on a cluster. The monitored data is injected into models to predict CPU time for different job sizes on the same or on a different cluster. Also, the complexities of algorithms with respect to CPU time, and number and sizes of the messages are estimated. When comparing these estimations to what one would expect, it is possible to pinpoint poor application efficiencies. This already led to application improvements on one node, and on a cluster. After correction, the codes showed perfect scaling with respect to the problem size, and to the number of processors used. Optimization of applications to reduce execution and communication times is clearly a very efficient path towards a better resource usage, and, as a consequence, to a reduction of the energy consumption per running job.

1.5 Application design

In the previous sections, it was supposed that the data is placed contiguously in main memory such that it can be accessed with stride 1 as in the paging mechanism. If an indirect addressing is needed, or the main memory is not accessed consecutively, another order of magnitude in performance is lost. Sometimes, it is possible to keep vectors in cache; cache misses can then be reduced, and performance increased. Nevertheless, this is not always possible. For instance, in real-life finite element applications, the vector lengths are very long, the matrices are sparse, and the resulting performance is small. Finite element methods could be replaced by spectral element methods [43]. Then, the polynomial degree can be high, the method converges exponentially for smooth problems, the number of variables is much smaller for the same precision, the matrices are less sparse, and efficiency is higher. Variables inside elements can directly be eliminated, the performance of the algorithm increases, the matrix condition of the remaining border variables is smaller, and the iterative matrix solver converges faster.

If one takes into account all those items, the F/J (Flops per Joule) ratio can be highly increased, the per job cost goes down, the waiting time is lower, and the work advances faster.

1.6 Organization of the book

1.6.1 Historical aspects

The book starts with a short history in high performance computing with an outlook to possible evolutions. We believe that it can be interesting to learn that many useful ideas have been around for a long time. Indeed, there are still researchers who were delighted with the old vector machines. They wrote simulation codes that were highly efficient in the usage of the limited amount of available main memory on the old vector machines like the Cray-1. Such legacy codes are still around and cause headaches when trying to port them to recent computer architectures.

Until the sixties, processors executed instructions in a fully sequential manner, thus using ten or more cycle periods to execute an add or a multiply operation. The applications that ran on those slow machines were coded sequentially. Mostly, one-dimensional problems were solved for which the complexity of the algorithms were not an important argument. This changed when pipelining and the simultaneous activation of several functional units brought the performance up. Parallel computing was pushed forward at the end of the nineties, reaching 106'496 nodes on the IBM Blue Gene/L machine at LLNL. One has the impression that a machine should not go over one million of nodes, and the additional performance increase must be realized by multi-processor, many-core nodes. One already talks about 1024 core nodes necessary to build an exaflop machine around 2018. Note that the graphical processor built by NVIDIA already had 240 cores in 2008. The evolution of hardware and software developments are sketched in this chapter with a special emphasis on the TOP500 list, and with a proposal to add new parameters to it.

1.6.2 Parameterization

To be able to adapt processor frequency to the needs of the application and to predict CPU and communication times to choose a well suited resource, the behavior of the application on the different computer architectures has to be described. For this purpose, the applications, the computers, and the internode communication networks are parameterized. There are parameters directly related to the application, the computer, and the network. Others describe how the applications behave on the computers and on the networks. Some parameters are important, others are rarely used. When a user talks about $V_a = 2$, one immediately knows that the application is main memory access driven. If $\Gamma < 1$ the application is dominated by internode communication, and another resource with a faster interconnection network having a $\Gamma >> 1$ has to be found. When $g_m = 0.3$, we know that the main memory access subsystem has a problem, and the underlying node architecture could be a NUMA machine with two or more processors and cores, or the main memory subsystem is new and not mature.

1.6.3 Models

Based on the parameterization the CPU, communication and latency times are estimated by means of models. These models predict complexities of CPU and communication times. They need information on relevant input parameters and measurements of previously executed jobs for which the timings and the input parameters have been stored in a data base.

1.6.4 Core optimization

A major energy reduction per job can be reached by optimizing the application on a single core. We have experienced performance improvements by factors typically between 2 and 5, just by applying the methods presented in the book, leading to high energy reduction for an executed job.

After a profiling of the application, the user can concentrate on the loops that dominate the CPU time. The first check is to control if data is stored with stride = 1. This means that data should be stored contiguously in main memory, i.e. profiting from the pipelined paging facility of the RISC computers. Then, some parameters have to be improved, and cache misses minimized. The procedure is presented with mathematical kernels such as the matrix*matrix multiplication BLAS3 algorithm, with a sparse matrix*vector kernel, and with real-life applications coming from plasma physics. To make execution acceleration visible, the compiler optimization levels are switched off. We shall see that compilers have reached a high level and well handle program optimization.

1.6.5 Node optimization

A node consists of a number of processors; each of which includes one or more cores. The node computer architecture can be a single core, a shared memory SMP machine, or a virtually shared memory NUMA architecture. The parallel node architectures can be programmed with OpenMP [89] interpreted by the compiler, by a multithreading approach, by an MPI library, or even by an automatic parallelization offered by the compiler that, at least at the moment, does not yet lead to a good scaling.

If OpenMP is used to parallelize shared memory or virtual shared memory architectures, the compiler is often unable to optimize since the transformations made by OpenMP become too complex. Then, it is advised to help the compiler by simplifying coding of loops in algorithms.

1.6.6 Cluster optimization

Optimizing on parallel distributed memory machines is done through the communication library MPI. Parallelizing codes is not a simple task, and should only be made when absolutely necessary. Typically, MPI parallelization is needed when the application needs more main memory than a node can offer, or if the execution time on one node takes days. A finite element example is used to sketch the parallelization procedure on a distributed memory computer architecture. Then, examples of parallelized programs are shown, the CFD (Computational Fluid Dynamic) spectral element code SpecuLOOS [45, 46] is discussed in details. We shall see that this code showed strange behaviors if more than 1000 cores are used. Applying the models described in the "models" chapter led to the detection of an improper implementation. After correcting it, the code perfectly scales up to 8192 cores. Again, improving an application, here a parallel application, reduces the execution time, and, as a consequence, the cost and the energy consumption per run go down.

1.6.7 Grid-brokering to save energy

A cost model is presented that is based on the CPU, communication and latency time prediction models presented in chapter 4, and on data stored on a data base. This data base includes parameters that describe the resources and data on the application is collected during execution by the specially developed monitoring system VAMOS [88]. All that data is necessary to calculate the costs for CPU, communication, latency, and waiting times, energy, licensing, and data transfer. The models are combined in a cost function, and realized within the Intelligent ApplicatioN-Oriented System (ïanos). Its goal is to submit the application to the most adapted resource, to the resource that consumes the least amount of energy, or to the machine that takes the smallest turn-around time. Such choices can be influenced by the choice of some free parameters.

At the end of this chapter, it is shown how a dual-core processor frequency of a laptop can be adapted to application needs. The temperature measured between the two cores is lowered by up to 30°. Since MTBF (Mean Time Between Failures) increases by a factor of 2 per 10° temperature reduction, not only does the battery lasts 30% longer, but the MTBF is 8 times larger.

The book ends with some recommendations, a Glossary and the References.

Chapter 2
Historical highlights

"Je lis dans l'avenir la raison du présent."
Alphonse de Lamartine, French writer (1790–1869)

Abstract This chapter acts as a short historical survey of both the computers and the applications evolution. The evolution of the computers starting from the "Bones" of J. Napier to compute the Logarithms in the XVIIth century to the last up-to-date IBM BG/Q supercomputer is first presented. We then parse the algorithms evolution from the second century BC to the Car-Parrinello method. The TOP500 and GREEN500 supercomputer lists are discussed and a new parameter is proposed that better characterizes main memory access dominated applications.

2.1 Evolution of computing

In the present chapter, we concentrate on information that can contribute to the understanding of the more recent hardware, to the basic software evolutions, and to a better design of HPC applications. Important dates and inventions are presented in Table 2.1. Further details on the evolution of the computing until 1965 has been described superbly in the book by Michael R. Williams [142], and a more in depth look into the detailed organization of the computers can be found in the book by Tanenbaum [126].

The Abacus [55] was very probably the first computing support. It is still intensively used in Asia. The invention of the logarithms initiated computing in Europe. The first table of logarithms was made by the Swiss mathematician and watch builder Jobst Bürgi who used it to support his private studies on mathematical series applied to astronomy. The natural base and 10 base logarithms were published by Napier and Briggs shortly afterwards. To compute the logarithms precisely, Napier built a mechanical computing machine using single-digit multiplication tables engraved on thin rods. Although Napier called these "numbering rods", the designation "Napier's bones" (see in [142] and [56]) became popular, as durable sets usually were made from animal bone, horn, or ivory. Blaise Pascal built an add and substract machine for his father who was a tax collector. Other mechanical

R. Gruber, V. Keller, *HPC@Green IT*,
DOI 10.1007/978-3-642-01789-6_2, © Springer-Verlag Berlin Heidelberg 2010

Table 2.1 The evolution of computing

Year	Who	Technology	Type/name	Application
1617	J. Napier	"Bones"	Multiplication	Logarithms
1623	W. Schickard	Mechanical	Add/substract	
1642	B. Pascal	Mechanical	Add/substract	Tax collection
1666	S. Morland	Mechanical	Add/substract	For sale
1674	G.W. Leibniz	Mechanical	All operations	
1678	R. Grillet	Mechanical	All operations	Fairy machine
1820	C. F. Gauss	Human	Least squares	Geodesy
1822	C.X. Thomas	Mechanical	All operations	Commercialised
1834	Ch. Babbage		Differences	Mathematical tables
1890	H. Hollerith	Punched cards	All operations	Accounting, census
1928	IBM	Punched cards	All operations	Business
1935	K. Zuse	EM relays	Z1	First computer
1943	A. Turing	Tubes, relays	Collossus	Code breaking
1946	Eckert/Mauchly	Tubes, relays	ENIAC	
1948	J. von Neumann	Tubes	IAS	Serial
1948	Bardeen, Brittain, Shockley	Transistor		Start of IC
1948	Williams, Kilburn	CRT	Mark I	First electronic computer
1949	Wilkes, Wheeler, Gill	CRT	EDSAC	Relocatable, linkable jobs
1968	Slotnick, NASA, Burrough	IC	ILLIAC IV	Array
1976	Seymour Cray	SSI	Cray-1	Vector
1997	Sterling et al.	Commodity	Beowulf	Cluster
2002	NEC	VLSI	NEC SX-6	Earth simulator

monsters mentioned in [142] were presented at fairs, or sold to wealthy people to impress their friends.

Almost 200 years after the invention of the logarithms and of the appearance of the first mechanical computers, Gauss came up with the least square algorithm in 1795. He used them in 1801 to calculate an accurate position of the asteroid Ceres, and to distribute optimally the triangularization errors appearing in the geodesy work at Hannover he accepted to manage from 1828 to 1847. The resulting system of linear equations were solved by a Gauss elimination procedure, also his invention. This systematic data processing method is at the origin of nowadays computer simulation programs. To resolve the big systems of linear equations, he definitely needed a big computational performance. Instead of using mechanical computers, Gauss [121] employed a number of persons who performed the matrix elimination process "by hand", i.e. using the logarithmic tables that Gauss redeveloped. One says that Gauss and some of his employees needed less than 10 seconds to perform a multiplication and an add. This was indeed the first practically applied high performance computing effort. Let us remark that the Gauss elimination algorithm has already been used approximately 2000 years before Gauss [30], when linear systems of up to 5 variables were solved in China. Gauss did not know about this work when he invented the method of least square fits [87].

After the difference engine [11], the analytical engine has been designed by Charles Babbage and programmed by Lord Byron's daughter Ada Lovelace. In fact, she was the first computer programmer. The programming language "Ada" was named after her. The machine was specialized to compute high derivatives of functions by means of consecutive differences. A prototype machine typically weighted 5 tons, was expensive to build, and never worked. This machine can be considered as a precursor of the mechanical arithmetic machines that were used until 1970. In 2002, a running replica of the *Difference Engine Number 2* has been constructed in London, and is now exposed in the Computer History Museum in Mountain View, Cal.

In 1890, Hermann Hollerith added punched card systems to mechanical engines to accelerate accounting for the American census of that year. The time gained to evaluate the large amount of data reduced the estimated costs by 5 M$. The Company that he founded after this success merged with two others to become IBM in 1924 that remained leader in the punched card systems for over 30 years.

One hundred years after Babbage, Konrad Zuse constructed the first automatically controlled calculating machine. He applied for a patent in 1936. The instructions to be executed were read from a punched tape, results were written on punched tapes. The data was stored in binary form. The operands to be handled by the arithmetic unit had two parts, the 16-bit mantissa, and the 7-bit exponent. Still today, a computer consists of an arithmetic unit, a memory based on a binary system, and an I/O system to read data in and write results out. The first prototype was destroyed during the Second World War, and the first productively running Zuse machine was the Z4 installed at ETH, Zurich in 1950. It produced results until 1960 when it was replaced by the ERMETH computer built at ETHZ by A. Speiser, E. Stiefel, H. Rutishauser, and A. Schai.

Parallel to Zuse's effort, John Atanasoff and Clifford Berry started at the Iowa State University to put together the ABC computer to solve large sets of equations. In 1942, the electronic calculator was close to working when Anatasoff joined a Navy laboratory, and his work became secret. The ABC machine influenced John Mauchly at Moore School in Philadelphia to develop an electronic computing machine, the ENIAC [72], and later the EDVAC [135] computers. John von Neumann from the Institute for Advanced Study (IAS) at Princeton, NJ published in 1945 the internal Moore School paper on *First draft of a report on the EDVAC computer*, published later in [135]. Most of the published material came from the design team of the ENIAC and EDVAC projects, J. Presper Eckert and John Mauchly, and probably also from Atanasoff and Berry. Thanks to this publication, a one processor serial computer having a control system, a processor, a memory, and a program stored in a random access memory is called a *von Neumann computer* (see Fig. 2.1).

At Harvard, with the partnership of IBM, the Harvard MARK machines were built under the leadership of Howard Aiken. One of Aiken's assistant was Grace Hopper who became famous because she wrote the first compiler that strongly influenced the Fortran and Cobol languages. She discovered the bug, a real moth, in a relay of the Harvard Mark II Relay Calculator, and later became an admiral in the

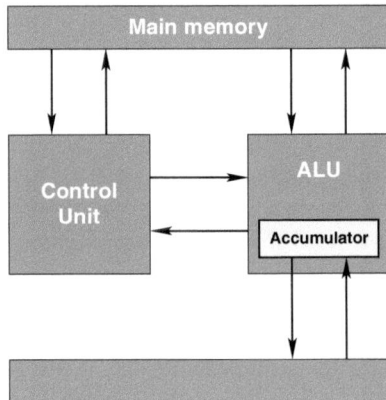

Fig. 2.1 The "von Neumann" computer architecture

US navy. George Stibitz designed the Complex Number Calculator at Bell Labs and John von Neumann built a computer at the IAS.

At the beginning of the Second World War, in Great Britain, the *Bombs* were built on the basis of machines developed in Poland to decipher the codes produced by the German Enigma machine [105, 142]. The Bombs were followed in 1943 by the much more advanced Colossus computer [105, 142] designed by M.H.A. Newman and built by T.H. Flowers. In Manchester and in Cambridge, the Ferranti Mark I based on CRT [140], MADAM [141], and the EDSAC [138] computers were constructed based on von Neumann's paper [135]. Whereas the ENIAC computer had to be rewired when changing a program, these machines were able to execute a sequence of instructions read from a paper tape memory system. The EDSAC machine included branching, and can be considered as the first productive von Neumann machine. The Ferranti Mark I machine was commercialized by Ferranti. It was sold to Toronto University and to several clients in UK. The next machine that was developed in common between Manchester University and Ferranti was the Atlas computer that included the virtual memory concept. The design proposal to build an interactive system was refused due to money shortage. Even though very innovative and successful in UK, the British computer scene led by Ferranti was not successful on the world market.

The array and cluster computer scene has been initiated by the ILLIAC IV array machine [12] built as an 8×8 mesh of processors. It was designed as a PDE (Partial Differential Equation) solving machine based on domain decomposition, each processor could take care of a subdomain. The machine was delivered in 1972 to NASA Ames, and offered a full service in 1975. The MPP (Massively Parallel Processor) concept was realized in the Staran machine by Goodyear Aerospace [14], by the Intel Paragon [40], by the Thinking Machines CM-5 [40], and by the Cray T3D [40] and Cray T3E [114]. Since 1997, the Beowulf concept [119] of massively parallel commodity processors interconnected by a Fast Ethernet or by a high bandwidth network such as Gigabit Ethernet (GbE), Myrinet (www.myri.com), Quadrics

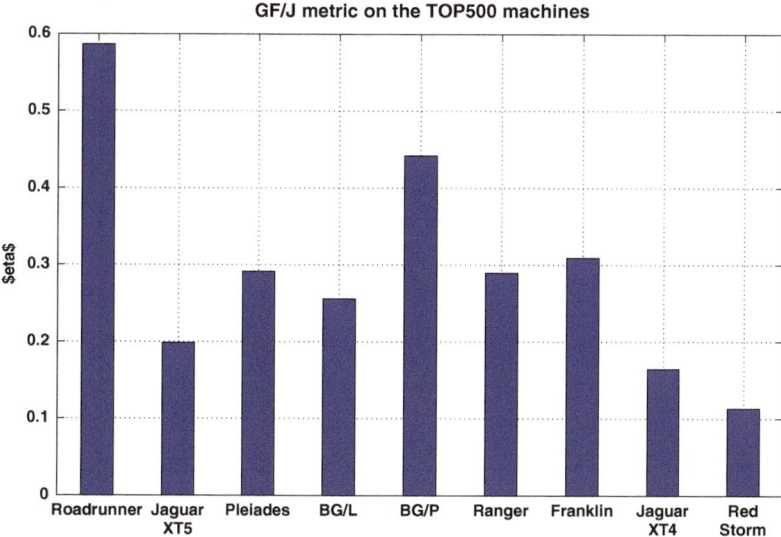

Fig. 2.2 Comparison of the Linpack η values (GF/J), i.e. the number of operations per consumed Joule, of the 9 top machines in the TOP500 list of November 2008. The number 1 machine (Road-Runner based on Cell BE processor at Los Alamos National Laboratory) is also the most energy efficent machine. The second most efficient machine is the number 5 TOP500 machine: a BG/P Solution with two quad-cores PowerPC 450 processors per node at Argonne National Laboratory. The difference between the number 1 and the number 2 (Jaguar is a Cray XT5 quad-cores Opteron processors based cluster) machines is 3 times in energy efficiency

(www.quadrics.com), or Infiniband (www.infinibandta.com) has become very popular because of its increasing high performance/price advantage.

Now, the evolution goes in the direction of high performance/Watt ratios (see Fig. 2.2). In fact, we propose to introduce the quantity η that measures the GF/s rate achieved with one Watt. This leads to the GF/J, i.e. the number of operations in GF that can be performed within 1 Joule of energy. The DSP and FPGA processors for embedded systems, the general purpose graphical process units (GPGPU), and the IBM Blue Gene machine tend in this direction.

Let us come back to the memory bandwidth that is a real performance bottleneck for most scientific applications. In fact, the bandwidth of the memory subsystem lags behind the processor speed. Two methods to increase memory bandwidth are used: memory banking and parallel access. In vector machines such as the Cray machines or the NEC SX series, the memory is cut into a number of banks needed to achieve a main memory bandwidth in words/cycle period comparable to the number of results produced by the processors. In 1975, the Cray-1 had 16 banks. For a NEC SX-5, with its 128 arithmetic pipelines per SMP unit of 8 processors, 16'384 memory banks were necessary to guarantee a memory bandwidth of 128 (64 bit) words per cycle period, corresponding to the maximum number of results per cycle period. The first commodity processor that incorporated four main memory banks was the Alpha machines from HP. Today, cost effective DDR dual access memory can deliver more than 10 GB/s memory bandwidth bringing the PC world to the HPC level of 20 years ago.

2.2 The first computer companies

2.2.1 ERA, EMCC and Univac

The Moore School projects ENIAC and EDVAC resulted in the Eckert-Mauchly Computer Corporation (EMCC) founded in 1946. EMCC built a computer for the US census office, and called it UNIVAC. The ERA Company grew out of WWII US Navy code-breaking group. Just before getting bankrupt, these companies were bought by Remington Rand, EMCC in 1950, and ERA in 1952, and became Univac, now Unisys.

Engineering Research Associates constructed in the early fifties the ERA 1100 series of computers for which Seymour Cray designed the control system. The 75 RePeat instruction of the ERA 1103A forms an early example of vector processing. Two of the chief engineers were Seymour Cray and James E. Thornton, later architect of the CDC STAR-100. Univac turned the ERA 1100 into the successful Univac 1100 series, machines working with 36 bit words. Seymour Cray worked for Univac as the responsible designer of the NTDS project for the Navy. He left Univac with other members of the NTDS team to join the CDC (Control Data Corporation) Company.

Univac entered again in the supercomputer business at the end of the fifties with the LARC machine constructed for the Lawrence Radiation Laboratory in Livermore.

2.2.2 Control Data Corporation, CDC

CDC was founded in 1957. William C. Norris was CEO for 30 years. Seymour Cray and James E. Thornton were the most eligible hardware and system software engineers. The first commercialized computer was the CDC 1604, a 48 bit word machine, delivered in 1960. This computer was a first commercial success and its performance exceptional, see Fig. 2.3. The CDC 3600 was delivered in 1963, followed by the CDC 6600 in 1965, a 60 bit word computer. This latter machine was much faster than what Moore's law predicted. This can be seen in Fig. 2.3. The 6600 was also the turning point in this Figure versus a much slower single processor performance increase depicted with the green lower line. Jim Thornton was the chief architect of the processor, whereas Seymour Cray took care of the I/O system, of the system software, and of the Fortran compiler. In a Foreword of a book by Jim Thornton on the CDC 6600 that appeared in 1970, Seymour Cray wrote: "This book describes one of the early machines attempting to explore parallelism in electrical structure without abandoning the serial structure of the computer programs. Yet to be explored are parallel machines with wholly new programming philosophies in which serial execution of a single program is abandoned."

After the CDC 7600 arrived on the market in 1969, James E. Thornton developed the STAR-100 and Seymour Cray left CDC in 1972 to form his own company Cray Research Inc. CDC was again commercially successful with the Cyber 200 series. The supercomputing branch was outsourced in 1983 when the ETA Systems Inc. was established to build the ETA10 machine. ETA was closed in 1989.

2.2.3 Cray Research

Seymour Cray was the founder of Cray Research in 1972. The first Cray-1 prototype was delivered in 1976 to LANL. It was built as a very high performance sequential machine. The first users realized that it was possible to run the two functional add and multiply units in parallel and totally pipelined. The vector machine was born.

In 1981, Steve Chen's Cray X-MP was the first SMP machine at Cray Research. It had four processors, all linked to the shared main memory subsystem. This machine was a commercial success. Seymour Cray and Steve Chen always closely listened to the users. When some important ones claimed that DAXPY ($\mathbf{y} = \mathbf{y} + a^*\mathbf{x}$, a is a scalar, $\mathbf{x}$ and $\mathbf{y}$ are vectors) was their most often used vector operation, the 8 processor SMP machine, called Cray Y-MP, a successor machine of the Cray X-MP, was able to execute this vector operation at top speed. Specifically, per cycle period, the Cray Y-MP was able to load two 64 bit words and store one. Since the multiply and add operations are pipelined and can operate simultaneously, one DAXPY result could be obtained every cycle period by each processor. In modern cache-based RISC architectures, one DAXPY operation typically takes 10 to 100 cycle periods. These SMP computers are well described in the book by Hockney and Jesshope [78].

Compilers were written to efficiently exploit vectorization, and even the SMP parallel architecture could be used in a user friendly manner through the micro-tasking or the auto-parallelization concept. Thus, the winner of the first Cray Gigaflop Performance Award in 1989 was able to parallelize his program in three weeks using the auto-tasking option and the help of a Cray Research engineer. The achieved measured sustained performance on an eight processor Cray Y-MP was 1.708 Gflop/s [6] with a speedup of 7.3 out of 8 and a sustained efficiency of 70%.

In parallel to Steve Chen, Seymour Cray designed the Cray-2 machine, a four processor SMP machine with 2 GB of main memory, arriving on the market in 1985. The cycle period was 4.1 ns, leading to a close to 2 GF/s peak performance. For main memory access dominated applications, in today's cluster designs a rule of thumb still is 1 GB of main memory per sustained GF/s.

Cray Research was purchased by SGI in 1996, and became Cray Inc. in 1999. It merged with Tera Computer Comp., and mainly commercializes now the Cray XT3 supercomputers.

Seymour Cray finished his career at Cray Computer Corp. he founded in 1989, where he built the Cray-3 and Cray-4 machines based on the GaAs technology. These machines could never be commercialized.

Seymour Cray did not only design and build high performance computers, but he was also deeply concerned with the software aspects that were integral parts of the computer. His goal was to construct highly equilibrated computers that could run at high efficiency for all HPC applications.

2.2.4 Thinking Machines Corporation

Thinking Machines Corporation (TMC) was a US supercomputer company that built the Thinking Machines around the idea developed in the PhD thesis of W. Daniel Hillis, a computer scientist at MIT [74]. Hillis starts his thesis by *"Someday, perhaps soon, we will build a machine that will be able to perform the functions of a human mind, a thinking machine"*. Outside the fact that Hillis was successful in the dream of every computer scientist (the Connection Machines passed the Turing test [132]), he designed supercomputers in a very different way.

The CM-1 machine

Hillis proposes a 64K processors supercomputer, all processors executing one instruction at the same time. The first machine of Thinking Machines Corp. was the CM-1. The CM-1 was not a success even if some of them were sold (for a price of 5 M$) under pressure of DARPA [127]. The CM-1 was built around RISC processors and a grid-topology interconnecting network. The machine was not able to process Fortran codes (it was designed to be programmed with a LISP dialect) nor could it perform floating point operations.

The CM-2 machine

This CM-1 failures were corrected in 1986 with the next machine, the CM-2. The CM-2 was also built around RISC processors, but interconnected by a hypercube-topology network. A Fortran compiler existed, but was not standard, and the users had to learn new programming techniques. Again, the CM-2 was not a success in the computational science community that mainly used Cray machines.

The CM-5 machine

In the late 1980's, with the end of the cold war, the US administration and DARPA reduced funding of military-purpose supercomputer projects, and concentrated more to the new "grand challenge" projects such as global climate modeling, protein folding, human genome mapping, or to earthquake predictions [28]. These projects require a high performance supercomputer. The goal was to build a 1 Teraflops supercomputer. TMC tried to achieve this goal with its last machine, the CM-5.

The CM-5 was designed to scale up to 16'384 computational nodes. One computational node was composed of one SPARC RISC processor, a vector unit and 32 MB of memory. One single computational node of the CM-5 was able to sustain 64 MF/s for a matrix-matrix multiply. The interconnection network was a Fat Tree. TMC estimated a performance of 700 GF/s on the LINPACK benchmark [75]. In June 1993, a CM-5 with 1024 computational nodes was number one of the TOP500 List. The TOP500 list lists the 500 fastest supercomputers every 6 months. The only metric is the Gauss elimination procedure of full matrices, called Linpack, or HPL (High Performance Linpack). More details can be found in Section 2.8.

The 1 TF/s goal was never achieved by a CM-5. The first supercomputer above 1 TF/s was the ASCI Red in June 1997 (according to the TOP500 List) based on Intel Paragons and installed at Sandia National Laboratory (SNL).

2.2.5 International Business Machines (IBM)

International Business Machines Corporation (IBM) is the oldest still active IT company, specialized in building computers for industry and business. IBM builds supercomputers more to follow the technological evolution than for profit, since the supercomputer market is too small. This slightly changed in 1997 with *Deep Blue*, the first computer ever to defeat a human chess world champion: the Russian International Grandmaster Garry Kasparov.

Before 1997

IBM had become wealthy with Hollerith's punched card machines to improve performance of business related jobs. Another important client was the American government to perform statistics. IBM was strongly involved in Harvard Mark I project in 1944. The Harvard Mark I computer had a decimal numerical system and performed one 23-digit decimal fixed point operation per second. This machine was made of 78 adding machines and calculators linked together, 765,000 parts, 3,300 relays, over 500 miles of wire and more than 175,000 connections [129]. It is considered as the first automatic digital calculator in the United States.

IBM also constructed the one-shot Naval Ordonance Research Calculator (NORC) machine for the US Navy installed in 1954. That machine was capable of 15'000 complete arithmetic operations per second. It has features such as floating-point operations, automatic address modification, and automatic checking of arithmetic accuracy.

In the early 1950's, IBM built the scientific 701 machine and the commercially successful 702, followed by 704, 705, and 709. The 701 is also the ancestor of the 7000 series which was available until 1968. The totally reengineered 360 series came on the market in 1964 and lasted until 1977, and was IBM's greatest financial success. The upward-compatible IBM System z^{10} is still available today.

Deep Blue shines

In 1997, IBM took the supercomputer world by surprise when announcing the *Deep Blue* [31] based on SP-2 machines, each consisting of 30 RS/6000 nodes. The clock was at 120 MHz, a node had 1 GB of main memory, the nodes were interconnected through a high-speed network. Apart of the common RS/6000 processors, special "chess chips" were added. The overall machine was able to perform 100'000 to 300'000 chess moves per second. The machine defeated Gary Kasparov $3\frac{1}{2} - 2\frac{1}{2}$ in May 1997.

The Blue Gene and Roadrunner era

After the arrival of the Japanese Earth Simulator (ES) in 2002 that was number one in the TOP500 list for two and a half years, DARPA launched a nationwide program to get back to first position. In 2004 IBM announced the Blue Gene/L (L for Linux) parallel computer based on low frequency PowerPC processors with a cycle period of 700 MHz, initially designed for embedded systems. Since energy consumption is proportional to frequency squared, for the same energy consumption, the number of nodes can be 16 times larger than for clusters with 2.8 GHz nodes. The arising networking problem was solved by superimposing a Torus network for point-to-point operations, a Fat Tree for multicast messages, a special network for barriers, and a GbE switch for administration. In 2005, the Blue Gene/L at Lawrence Livermore National Laboratories became number one with 32'768 dual core nodes, and the biggest Blue Gene/L at Lawrence Berkeley National Laboratory had 212'992 cores in 2008. The newer Blue Gene/P (P for PF/s) runs at 800 MHz, and has 4 cores per node. The largest installed P-system in the world is at Jülich Forschung Zentrum in Germany with 72 full racks (72'000 processors) with over 1 PF/s for the HPL benchmark. The next generation of Blue Gene, the Q-series ("Q" is after "P") targets a performance of 10 PF on the LINPACK benchmark in 2010/2011.

The first *petaflops machine* (10^{15} floating point operations per second) on the HPL benchmark was the *Roadrunner* installed at Los Alamos National Laboratory in 2008. It is based on Cell nodes, Fig. 6.6, designed by IBM for Sony's PlayStation 3 game console.

2.2.6 The ASCI effort

The successful Japanese machines that appeared in the nineties described below activated a strong American reaction with the "colored" ASCI program, first ASCI Red at SNL (June 97 to June 00) with up to 9'632 Intel cores, then ASCI White (November 00 to November 01) at Lawrence Livermore National Laboratory with 8'192 IBM Power 3 cores, and an HPL performance of 7.226 TF/s.

2.2.7 The Japanese efforts

The major players in Japan are Fujitsu, Hitachi, and NEC. These companies mainly produced computers that were compatible with IBM machines to penetrate the business and industrial IT market. However, there were several efforts made to achieve the first place in the TOP500 list.

Fujitsu

In 1993, Fujitsu together with the Japanese National Aerospace Laboratory finished the design of the (numerical) Wind Tunnel and reached number 1 in the TOP500 list in November 1993. The machine consisted of 140 Fujitsu vector computers with an HPL rate of 124 GF/s. One year later, after having lost the first rank to Intel for half a year, an improved version of the Wind Tunnel with 170 GF/s reached again the top for another one and a half years.

Hitachi

Then, Hitachi replaced Fujitsu at the top for one year with a pseudo vector machine of 2048 processors. This machine was installed at the Tsukuba research center.

NEC

The biggest Japanese success was certainly the Earth Simulator installed at Tokyo University. This installation in June 2002 was a real shock for America's supercomputing community since it reached 35.86 TF/s with HPL, 5 times faster than America's number one. The ES consisted of 640 eight-core NEC SX-6 supercomputers, each core delivering 8 GF/s peak. With this HPL performance, the ES remained number one until the end of 2004. Then the Blue Gene/L took over until the end of 2008. However, we have to mention here that for many main memory access dominated scientific applications the ES performance is higher valuable than that of the most recent TOP10 supercomputers. The NEC SX-9 that appeared in 2008 can deliver up to 100 GF/s. The NEC SX series was designed by Tadashi Watanabe, winner of the Seymour Cray prize in 2006 distributed in US to most notable computer architects.

Just by comparison, the total main memory bandwidth of the ES corresponds to 320 TB/s, the same as the total main memory bandwidth of the 131'072 cores of the Blue Gene/L at LANL (number one from November 2004 until June 2007). Since the efficiency of the main memory subsystem in vector computers is better than in most RISC architectures (Itanium is an exception), the ES would have been faster than the Blue Gene/L at LLNL for main memory access dominated applications. This is also true when comparing ES with the Roadrunner at LANL, number one since 2008. For matrix times vector (BLAS2) operations, NEC's ES would have probably been number one until 2009.

2.3 The computer generations

The computer technology started with relays. Vacuum tubes were at the origin of what one calls the first generation computers. With the invention of the transistors, the computer technology evolved very rapidly. The second generation machines used single transistors as constructing elements. Then, the circuits were integrated (IC), largely integrated (LSI), then ultra largely (VLSI) integrated. By the miniaturization of the technology and the multilayer concept, the density of circuits per chip continuously grows (ULSI and VHSIC).

To increase computational performance a cluster consists of many computer nodes; each node has a number of processors with many cores. A set of such parallel machines is now called a computational Grid [52, 53].

2.4 The evolution in computing performance

The lower line segments in Fig. 2.3 show the evolution of the single processor machines. One recognizes that the performance follows Moore's law (a four fold increase every 3 years, he stated this law with respect to the increase of transistors per chip, but it also predicts the performance of HPC machines) between 1947 and 1974, whereas from 1965 to 2002, the increase in single processor performance has only doubled every three years. Since 2002, the frequency of the processors did not evolve and stagnated at 3 GHz and less.

Fig. 2.4 shows a performance comparison between vector computers and desktop machines, or co-processors to accelerate desktop machines. In 1980, the most advanced Cray X-MP processor was 10'000 times faster than the Motorola 8087 co-processors. As one can see in the Fig. 2.4, the arrival of the commodity RISC processors in 1989 rapidly changed the image. In 2004, an Alpha EV7 or a Pentium

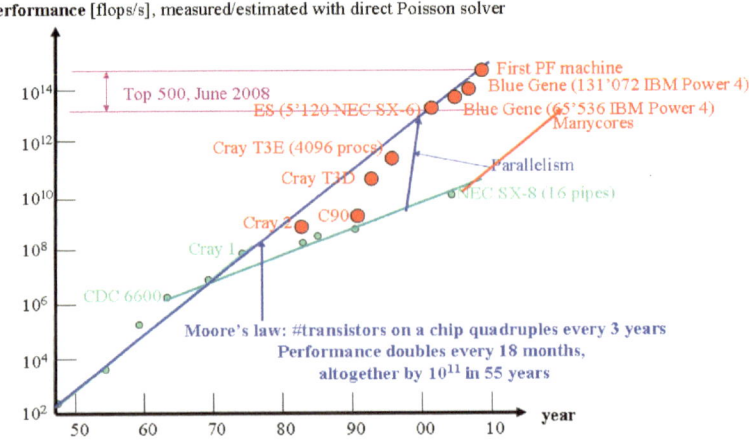

Fig. 2.3 Performance evolution of parallel computers

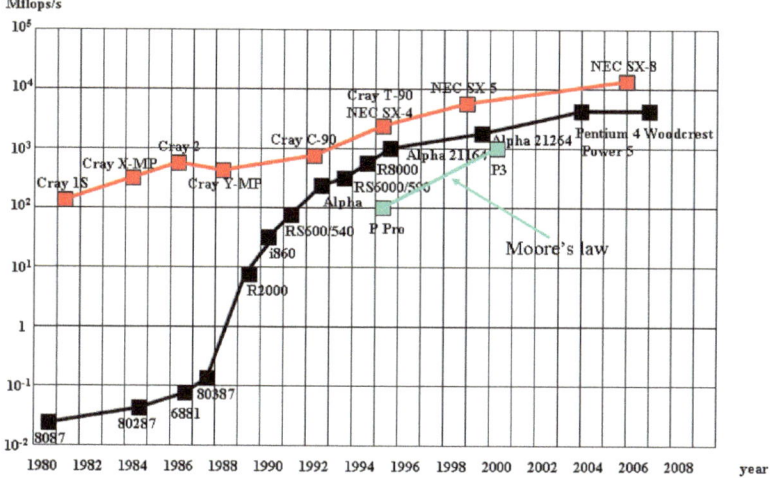

Fig. 2.4 Comparison between peak performances of vector computers and desktop machines

4 is only three times slower on average than the NEC SX-6 processor with 16 arithmetic pipes, and the recent commodity processors AMD Opteron, the Power 5 from IBM, and the Intel Itanium with very high bandwidth memory accesses came even closer to the vector processor line.

The overall peak performance per machine has followed Moore's law. The lack in single processor performance since 1974 has been compensated by a continuous increase of the parallelism (see Table 2.2 and Fig. 2.3), and by an increase in the number of cores since 2005.

Table 2.2 BLAS2 dominated Poisson solver performance evolution of some TOP 1 machines. *Porting this application on a Cell BE processor-based architecture has not been done so far

Machine	Year	No of proc	Performance	Unit
EDVAC (Univac)	1947	1	200	F
IBM 701	1952	1	3	kF
IBM 7090	1960	1	300	kF
CDC 6600	1964	1	2	MF
CDC 7600	1969	1	5	MF
Cray 1	1976	1	50	MF
Cray 2	1985	4	600	MF
Cray Y-MP	1988	8	1.5	GF
Cray C90	1992	16	8	GF
Intel Paragon	1992	512	30	GF
CM-5	1993	1024	60	GF
Cray T3E	1996	1024	150	GF
ASCI Red (Intel)	1999	9632	1	TF
ASCI White (IBM)	2000	8192	3	TF
NEC SX-6	2002	640	20	TF
IBM Blue Gene	2006	130'000	20	TF
Roadrunner	2008	20'000	*	TF

Let us make a remark here on the evolution of the main memory costs. In 1972, CDC sold a fast external memory costing 1$ per Byte. Thirty-three years later, for the same 1$ one could get 10 MB (MBytes). One realizes that main memory costs follow closely Moore's law.

2.5 Performance/price evolution

In the best case, the Harvard Mark I could execute one 23 decimal operation in about one second. It consisted of an instruction read, an operation on two ready-to-use registers, and a write back of the result, each step taking 0.3 seconds. The hardware and construction costs were estimated to 500'000$ in 1944. Thus, without taking into account the devaluation of the $ during 60 years, the cost per MF/s was about 10^{12} $ in 1944 (Mark I) and 1 $ in 2004 (Pentium 4). This is a 10^{12} performance/price improvement in 60 years, corresponding to a factor of 4 every 3 years. This exactly corresponds to Moore's law.

2.6 Evolution of basic software

The most striking basic software inventions are shown in Table 2.3. Ada Lovelace first programmed Babbage's Analytic Engine. Zuse invented an algorithmic language that he named Plankalkül. Grace Hopper wrote the first compiler, the Flowmatic, that strongly influenced the Fortran [104] and Cobol [120] compilers, both still in use today. Many languages have been invented during the 50 years of electronic computing history. The first parallel language was APL [85], Niklaus Wirth introduced Pascal [144], a follow-up of the not so successful Algol [108] language, and the object-oriented, modular programming Modula-2 [143] and Oberon

Table 2.3 Evolution of basic software

Year	Who	Basic software tools
1850	Ada Lovelace	Programmable difference schemes
1945	K. Zuse	Plankalkül (algorithmic language)
1954	Grace Hopper	Flowmatic (first compiler)
1955	J. Backus	Fortran
1959	Task Force	Cobol
1962	K.E. Iverson	APL (parallel language)
1971	N. Wirth	Pascal (structured programming)
1972	D. Ritchie	C
1978	Ritchie, Thompson	Unix
1978	N. Wirth	Modula-2 (object oriented compiler)
1979	J. Ichbiah	Ada
1983	B. Stroustrup	C++
1992	L. Thorvald	Linux
1990	V.S. Sundaram	PVM
1995	MPI Forum	MPI
1995	Sun	Java

languages. The American government tried to impose the new Ada [13] programming language, but was not successful.

Each computer manufacturer developed and sold its own system until 1990. Since then, two pseudo-standard systems have been adopted by most of the users: Windows from Microsoft for the PCs, and Unix [111] and Linux [125] for the servers. Together with Unix, the languages C [90] and C++ [122] have been developed and accepted by many users.

Since the arrival of parallel machines, different message passing libraries have been proposed to the programmers. The Occam library [102] was developed for the Transputer-based parallel machines that were popular in the eighties, and Intel proposed a new parallel language for the Paragon. Then, PVM [124] could be used on almost all the parallel platforms and was used in 2009 in about 20% of the scientific applications. Today, MPI (Message Passing Interface) [60] has become the standard message passing library that can be found on all the parallel machines. It is now possible to write portable parallel software by using MPI to transfer data from one processor to another one, even if this other processor comes from another manufacturer. SUN proposed Java [18]. A program has to be compiled once on one computer. The resulting relocatable can then be executed on other parallel platforms.

In this book, we will present examples written in Fortran 90, use OpenMP in shared memory parallel architectures, and MPI to pass messages between processors in distributed memory clusters.

2.7 Evolution of algorithmic complexity

Most of algorithms (see Table 2.4) applied today have been developed during the ramp-up phase of the electronic computing period. Gauss with the least square fits and his matrix elimination technique, and Jacobi [86] with his iterative solving method are exceptions. In fact, Gauss needed to develop his methods for the geodesic work in Hannover. Due to these methods, his maps were the most precise ones in the first half of the 19th century.

The arrival of programmable computing engines brought up a whole activity in the development of new algorithms adapted to solve large problems. The finite element method [36] and linear programming [41] were invented, the conjugate gradients [73], the Lanczos [92], the matrix diagonalizing LR or QR [112], and Multigrid [26, 70] algorithms are iterative methods to find solutions without destroying the sparse matrix patterns. The Householder transformation [79] transforms a dense matrix into a sparse one, and the FFT algorithm [35] reduces the complexity of the Fourier transform. The Monte Carlo [106], molecular dynamics [3], and Car-Parrinello [33] methods have been invented, and used in parallel computing.

Due to these algorithmic inventions, the computational complexity of the most efficient three dimensional Poisson solver has evolved [43] impressively. Poisson's equation

$$\Delta \Psi(\mathbf{x}) = f(\mathbf{x}), \tag{2.1}$$

Table 2.4 Evolution of algorithms

Year	Who	Method
-5th b.C	Pythagoras, Thales	Geometry
900	Al-Khwarizmi	Algorithms
1000	Persia/Irak	Algebra
1610	J. Bürgi	Logarithms
1614	J. Napier	Natural logarithms
1624	H. Briggs	Logarithms base 10
1808	J. B. J. Fourier	Fourier transform
1820	C. F. Gauss	Gauss elimination/Least squares
1845	C.G.J. Jacobi	Jacobi method
1943	R. Courant	Triangular finite elements
1946	Ulam, Metropolis	Monte Carlo methods
1950	C. Lanczos	Lanczos method
1952	Stiefel, Hestenes	Conjugate gradients
1958	H. Rutishauser	LR
1958	A.S. Householder	Householder transformation
1959	B.J. Alder	Molecular dynamics
1963	G.B. Dantzig	Linear programming
1965	Cooley, Tukey	Fast Fourier Transform (FFT)
1973	A. Brandt	Multigrid
1985	Car, Parrinello	Car Parrinello Molecular Dynamics

constrained with boundary conditions is relevant in many physics problems, for instance to evaluate the thermal conduction, the electrostatic potential, the diffusion in chemistry or neutronics, the pressure in fluid mechanics, Darcy's equation for porous media, Schrödinger's equation, the MHD equilibrium in plasma physics, the solidification of material, or when computing an adaptive mesh distribution. In the complexity estimations, tri-linear finite elements were used to approximate the solution $\Psi(\mathbf{x})$ of the Poisson equation in a three-dimensional space, N intervals have been chosen in each direction, altogether N^3 mesh cells. Results are given in Table 2.5. The complexity of the Gauss elimination process is given by $N^7 = N^3 * (N^2)^2$, where N^3 is the matrix size, and N^2 the matrix band width. With the Gauss-Seidel method, the number of operations is determined by the number of non-zero matrix elements $(27N^3)$ times the number of iterations (N^2), altogether $27N^5$. This method immediately eliminates the local, short wavelength errors, but the long wavelength errors are attenuated very slowly. To reduce the long wavelength error, an optimal successive over-relaxation (SOR) can be applied. Then, the number of iteration steps reduces from N^2 to $Nlog(N)$. As a consequence, the complexity reduces to $27N^4log(N)$. If the domain is cubic, a Fast Fourier Transform [35], or finite differences with a cycle reduction procedure [77] can be applied. Then, the number of operations reduces to $8N^3log(N)$. A real breakthrough was achieved with the Multigrid method [26, 70] in which the long wavelength errors are eliminated in coarse meshes, and fine wavelength errors in fine meshes. The computing time is proportional to the number of mesh cells. In fact, the number of operations is given by $270N^3$, where the number of multigrid V cycles has been estimated to 7. The Multigrid algorithm can be parallelized and the complexity is divided $N_{CPU}/log(N)$, where N_{CPU} is the number of processors used. This

Table 2.5 The evolution in algorithm complexity to solve Poisson's equation

Algorithm	Year	No of operations
Gaussian elimination	1947	N^7
Gauss-Seidel	1954	$27N^5$
Optimal SOR Iteration	1960	$27N^4(logN)$
Fast Fourier Transform	1965	$8N^3(logN)$
Cyclic Reduction	1970	$8N^3(logN)$
Multigrid	1978	$60N^3$
Parallel Multigrid	1981	$60N^3(logN)/N_{CPU}$
Parallel Gaussian	1985	$4N^7/N_{CPU}$
Improved parallel Multigrid	1989	$30N^3(logN)/N_{CPU}$

has to be compared with the direct Gauss elimination method whose performance increase is limited by a quarter of the number of processors. All these algorithms are described in [58, 113].

The overall gain in complexity from Gauss elimination to optimal parallel Multigrid is $N_{CPU}N^4/100log(N)$. For $N = 100$ intervals in the three directions, and $N_{CPU}=1000$, the gain is 10^9. With the computer performance improvement of 10^6, the overall increase in efficiency during 50 years gives an impressive 10^{15} factor, six orders of magnitudes coming from the performance increase of single processors, 3 orders from parallelism, and 6 orders from algorithmic improvements.

A similar evolution has been made when computing the ideal magnetohydrodynamic (MHD) stability behavior of thermonuclear fusion experiments such as axisymmetric Tokamaks or helically symmetric Stellarators. In 1972, the level of efficiency in eigenvalue solvers was such that on a CDC 6600 it was possible to solve the stability behavior of a one-dimensional (1D), infinitely long, cylindrical plasma [10] in one hour CPU time. In 1978, in one hour CPU time, the most unstable eigenvalue of a 2D toroidal Tokamak plasma [131] was computed on a Cray-1. To achieve a result with sufficient precision, it was necessary to switch to curvilinear flux coordinates, to a set of dependent variables adapted to analytic behaviors of the unstable solution, and to a sparse matrix eigenvalue solver, specially written for this purpose. The solution was approximated by a special finite element approach [61, 66], now called the COOL method [1].

In 1988, a 3D program has been written to study the stability behavior of helically symmetric Stellarator plasmas. The computing time of one eigenvalue is now less than one hour on an Intel Pentium 4, the same as on the Cray-2 and the NEC SX-3. Here, a special coordinate system [21], and special dependent variables have been chosen. The solution is approximated by a double Fourier expansion in the toroidal and poloidal angles, and by a COOL finite element approach in the radial direction. The eigenvalue solver has been optimized to the specific matrix structure [5].

2.8 The TOP500 list

Since 1990, a list has been assembled that contains the 500 most powerful computers in the world. Originators were Hans Meuer and Erich Strohmaier. They looked

for a single valued characterization of the computers. Jack Dongarra delivered the
Linpack benchmark.

The TOP500 list is published on the web (http://www.top500.org) twice a year.
The list is based on performance measurements made with the parallel Linpack
library giving a F/s (Flops per second) rate that is close to peak. This benchmark
is now called HPL (High Performance Linpack), and is dominated by matrix times
matrix (BLAS3) operations. The Linpack test case for the Earth Simulator (ES),
number 1 from 2002-2004, was made with a matrix of rank 1'041'216 needing
altogether 8.7 TB of main memory and 5.8 hours to run. This computation led to a
sustained performance of 35 TF/s out of 41 TF/s peak, this 85% of peak.

The tremendous evolution of the most powerful machine is shown in Figure 2.5.
A BLAS3 computation that would have taken one year on a Cray-1S processor with
an imaginary non existent huge main memory, needed about 4 seconds on the first
PF/s machine in 2008

When the TOP500 list appeared, the major part of the machines were installed in
research, governmental institutions, or used for weather prediction. Today, half of
the machines are in industry/economy, used for instance for semiconductor simulation, oil exploration, communication, car and airplane design, used in digital imag-
ing, banking, business, bioinformatics, and food industry. In 1990, the Cray Y-MP
consisted of 8 vector processors. This parallel, shared memory vector computer was
entirely constructed by the Cray company. They built the processors, the memories,
the memory switches, the I/O controllers, the operating system, the compilers,
everything. The machine was perfectly adapted to the specific needs of the HPC user
community. The arrival of high performance, low cost mass produced processors
and memories almost entirely stopped production of such machines. Cray and NEC

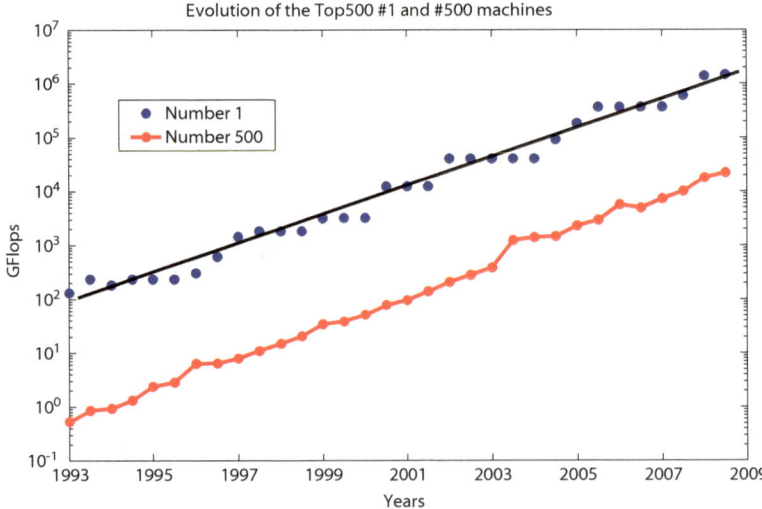

Fig. 2.5 Evolution of number one and the number 500 machines in the TOP500 list. The solid
black line shows double performance every year

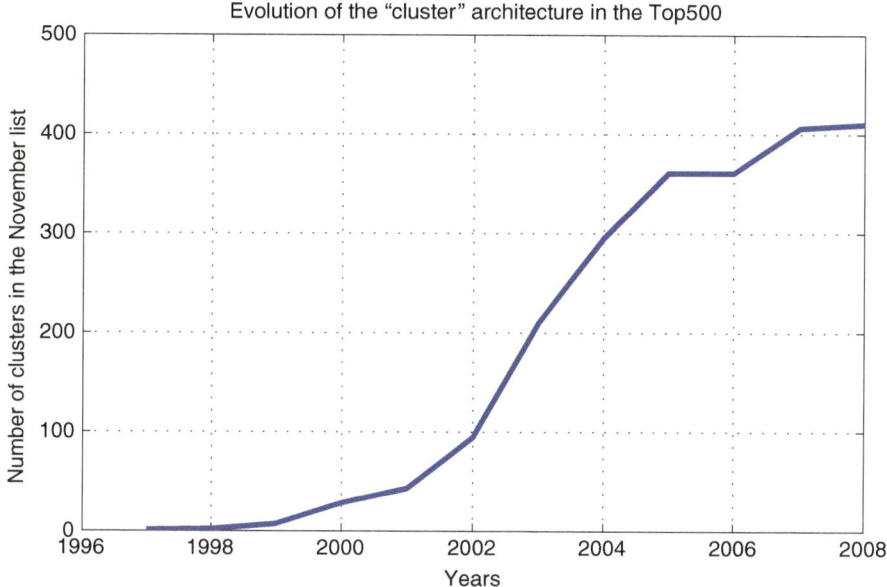

Fig. 2.6 Evolution of the cluster architecture in the TOP500 list

continue to build vector machines. The other remaining successful computer vendors switched to MPP and cluster concepts that include mass produced commodity
components, such as processors from Intel, AMD, or IBM, cheap memories, and
standard communication systems such as Fast Ethernet, GbE (Gigabit Ethernet),
Myrinet, Quadrics, or Infiniband. The basic software has also been standardised,
Linux, ScalaPACK, and MPI are such examples.

When looking at the type of processors installed in the TOP500 machines, it is
striking how the cluster architectures have evolved, Fig. 2.6. Since 1998, all the
TOP10 machines have more than 5000 processors. Starting in 2006, there is a new
trend towards special purpose HPC machines. IBM produces the Blue Gene series
with processors running at 800 MHz with a low energy per performance consumption. IBM also installed the first PF/s machine made of an AMD Opteron basis with
attached Cell BE nodes. The dominator of the graphics processor market, NVIDIA,
proposes now a new HPC node with four graphical processors including 960 cores
each. The machine has to be coded with the CUDA language, new to the HPC
community.

2.8.1 Outlook with the TOP500 curves

Figure 2.7 can be used to make some predictions on the evolution of the number one
machines, Fig. 2.5, the tenth in the TOP500 list, of number 500, i.e. the last one in

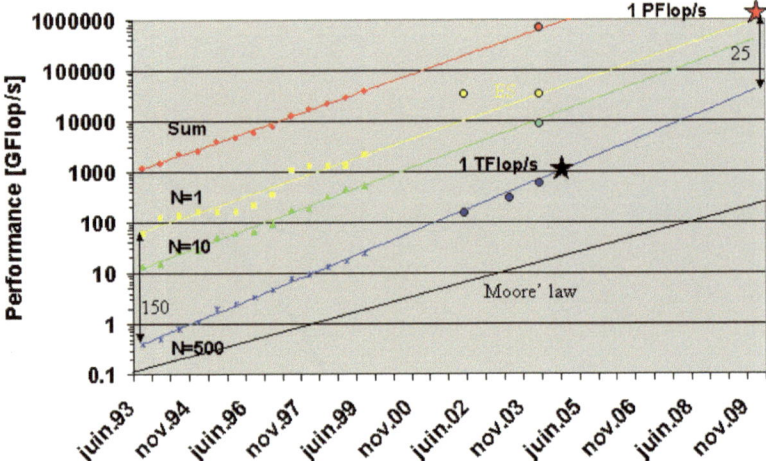

Fig. 2.7 TOP500 evolution

the list, and of the sum of the performances of all the 500 machines. In 2005, all the 500 machines in the list have passed the TF/s performance. Extrapolating the line for the TOP1 machine, the PF/s limit should have been achieved in 2010. In reality, the PF/s limit has been passed in March 2008. The machine is a hybrid cluster made of 6'912 dual-core Opteron processors from AMD and 12'960 IBM Cell eDP accelerators, each one consists of 9 cores running at 3.2 GF/s and producing up to 4 floating point operations per cycle period. Thus, the Cell eDP part of the Roadrunner installed in Los Alamos has 116'640 cores of 12.8 GF/s each.

The performance ratio between the first machine and machine number 500 (see Figure 2.7) was 150 in 1993, in 2010 this factor had shrunk to 25. This effect can be explained with the increasing number of low cost HPC installations, especially in industry/economy.

The evolution in the TOP500 of the number of clusters, or networks of work-stations and PCs (NOW) is depicted in Figure 2.6. One sees that the first cluster entered in the list when the Beowulf concept was published [119] in 1999. In 2002, the number of clusters reached 100, and was close to 300 in 2004. These low cost machines have stable software environments for resource management, monitoring, scheduling, controlling, and message passing. The management of such machines is no longer more complex than the management of vendor specific MPP machines.

2.8.2 The GREEN500 List

The choice of a metric that expresses the ecological cost of a supercomputer is not trivial. In the past, metrics such as MTBI (mean time before interrupt), MTBF (mean time before failure) or MTTR (mean time to restore) has been used to measure the reliability of a system but never to classify it. During the years of the "GHz race", processors became more and more subject to failure which is a crucial issue in the case of a multi-processor system. For very large systems (hundreds of thousands process units), MTBF of several hours have been observed. Chip founders lowered the frequency of their processors and added multiple cores in the same time, keeping the same performance for Grand Public computing. But, in the world of high performance computing, the only accepted metric is the pure CPU performance in F/s (TOP500 list) of the high performance Linpack (HPL), always reaching close to peak performance. A list with numbers corresponding to 70% of peak performance would be very close to the present one related to HPL. Thus, a high-end system built with processors with a lower peak performance would automatically fall down in the list, even though it would deliver the same performance for all applications dominated by main memory access, and consume less energy. In 2003, a new metric, the GREEN500 list [48, 115], to classify the supercomputers was proposed.

Since the June 2008 release, the TOP500 list also includes the power consumption of the listed machines measured during running the HPL benchmark. The power measurement is valid for all the *essential hardware, shared fans, power supplies in enclosures or racks*. It excludes "non-essentials" components in the room such as non-essential disks, air- or water-cooling jackets around air-cooled racks or UPS systems. It leads to measurements between 1.32 MW for a TOP10 system (248 MF/W) to 257 KW for a TOP500 system (122 MF/W). The number one machine (the Cell BE based Roadrunner at Los Alamos) consumes 2.35 MW.

The GREEN500 list measures "performance per Watt" or "number of operations per Joule" indicated as

$$\eta = \frac{\text{Number of operations}}{1 \text{ Joule}}, \tag{2.2}$$

η measures the number of HPL operations that can be performed during the time the machine consumes one Joule of energy.

In November 2008, the number one of the GREEN500 list was a Cell BE based cluster with $\eta = 0.536$ GF/J. That machine is ranked number 220 on the TOP500 list. Roadrunner, number one of the TOP500 list, is ranked 5 on the GREEN500 list with $\eta = 0.458$ GF/J. Note that IBM trusts the 20 first ranks. Number 1 to number 7 machines are clusters made of Cell BE, while number 8 to number 17 are Blue-Gene/P machines. This list definitely does not consider usefulness of the machines. For instance, the Cell BE nodes of Roadrunner, number one in 2009, are difficult to use, and, only a few real-life applications can profit from the PF/s performance, and if ever, only a portion of it.

In comparison, the Earth Simulator (installed in 2002) has a peak performance of $R_\infty = 40.96$ TF/s for a sustained performance on HPL of $R_a = 35.86$ TF/s, and was number one for five TOP500 list releases. It was ranked 73 on the November 2008 list, and consumed a total power of 6.8 MW, leading to $\eta = 0.0052$ GF/J for the HPL benchmark. This value is hundred times away from Cell based machines. But for BLAS2 dominated operations the ES is hundred times more efficient, and the η values would become comparable. This fact introduces the next subsection.

2.8.3 Proposal for a REAL500 list

The GREEN500 list inventors make the same assumption than those of the TOP500 list: the pure CPU performance based on the BLAS3 HPL benchmark is considered to be enough to classify the supercomputers. Despite of its simplicity, HPL encourages constructors to push the peak performance rather the "useful" one. For instance, compiled codes instead of assembler written (BLAS3) kernels can reach at most one third of the announced performance, and reach somewhat 1 to 10% of peak if the algorithm is main memory access bound (BLAS2). As we have remarked earlier, if the lists would be based on the results of a BLAS2 operation, they could be quite different.

We therefore propose to introduce a new REAL500 list including the following measurements:

1. HPL benchmark as in the present TOP500 list
2. A benchmark that measures the real main memory bandwidth by a benchmark such as SMXV (Sparse Matrix times Vector multiply)
3. The total energy consumption should include the power consumption of the machine and its cooling. The η value, in GF/J, should be given for the HPL and SMXV benchmarks.

Chapter 3
Parameterization

> *"Everyone has the obligation to ponder well his own specific traits of character. He must also regulate them adequately and not wonder whether someone else's traits might suit him better. The more definitely his own a man's character is, the better it fits him."*
>
> *Cicero,* Roman philosopher

Abstract The behavior of an application can vary depending on which resource it runs. To find a well adapted resource the applications and the resources are parameterized. We start with some definitions. Parameters are then defined to characterize applications, nodes, and networks. The behavior of applications running on nodes and networks are also described through a set of quantities, some of them are then used to predict complexities (chapter 4), to optimize applications, algorithms or kernels on cores (chapter 5), on nodes (chapter 6), on clusters (chapter 7), and to evaluate the costs of a submission (chapter 8).

3.1 Definitions

The IT literature describes a Grid as a multi-levels abstraction; we present it in a top-down manner. Starting from the highest hardware level (the Grid as level 0) to the lowest (a computational core as level 5), all the elements constituting the ecosystem of applications, resources, networks, and users are defined.

Let us give some definitions that will be used along the book. The context of the work is an HPC Grid defined as:

Definition 1 *an* **HPC Grid** *is a set $\mathcal{G}$ of sites (see Def 2) s_ℓ, $\ell = 1, \ldots, N_{site}$, geographically distributed and interconnected with standard, open, general-purpose network protocols such as TCP/IP over the conventional Internet. The Grid is operated by a middleware (between local schedulers and Applications) $\mathcal{O}$ such as Globus or Unicore. All HPC applications $\mathcal{A}$ running on this Grid and the users $\mathcal{U}$ (see Def 17) form the Grid community.*

> $\mathcal{G}$ *is the Hardware Level 0.*
> $\mathcal{A}$ *is the Application Level 0.*
> $\mathcal{O}$ *is the Basic Software Level 0.*

R. Gruber, V. Keller, *HPC@Green IT*,
DOI 10.1007/978-3-642-01789-6_3, © Springer-Verlag Berlin Heidelberg 2010

An HPC Grid is composed of a set of sites with this definition:

Definition 2 *a* **site** s_ℓ *is a set of resources administered and managed in a local manner by one administrator (or administration). Each site can have its special policy of usage.*

s_ℓ is the Hardware Level 1.

A site is composed of a set of resources:

Definition 3 *an* **HPC resource** *(or simply* **resource***) is a homogeneous computing resource r_i such as a workstation, a commodity cluster, a NUMA machine, or a supercomputer belonging to a Grid site s_ℓ, and having P_i nodes.*

r_i is the Hardware Level 2.

A resource is connected to the Internet and accessible by the authorized users. It is administered by a local administrator. An administrator takes care of one or several HPC resources by applying local usage policies.

Definition 4 *a* **computational node** *(or simply a* **node***) is a computational unit that is connected to the internode communication network. It can be an SMP machine, a NUMA machine, a workstation, a PlayStation3 (PS3), a graphical processing unit (GPU), or a laptop. A node has a power consumption of J_i [W]. A node has N_{CPU} processors.*

A node is the Hardware Level 3.

Definition 5 *a* **CPU** *(or a* **processor***) is defined by its number of cores (see Definition 6) N_{core}, and by a local cache memory that is connected to the main memory.*

A processor is the Hardware Level 4.

Definition 6 *a* **core** *is part of a processor. It contains at least one ALU (Arithmetic Logic Unit), registers and a local memory (level 1 cache memory). It is defined by its addressing mode am_{core} (32 or 64 bits), its frequency F_{core} [GHz], and by the number of double precision floating point operations ω_{core} it can perform during one clock cycle (number of functional units), and its efficiency g_p.*

A core is the Hardware Level 5.

There are cores with one or more computational units, a main unit, and an attached processing unit. In nowadays processors, the main unit has several functional units (one or multiple add and multiply units) that can be accessed concurrently. It can also include an additional attached processing unit. In the case of an Intel node this additional hardware is called SSE (for **S**treaming **SIMD E**xtensions [80]) that extends the x86 instruction set. SSE instructions use 16 special 128 bits registers able to store two double precision floating point operands that can be processed during one clock period. This leads to up to four double precision operations per cycle period (two adds and two multiplies). To access these units, library routines have to be called. The programmer or the compiler must use the special SSE instructions.

Definition 7 *the* **memory** *of a node is defined by its amount m_{node}, its peak main memory bandwidth M_∞ [Gw] (where a word (w) = 64 bits), its latency L_m [s], and its efficiency g_m. It can be connected to a node or to a processor.*

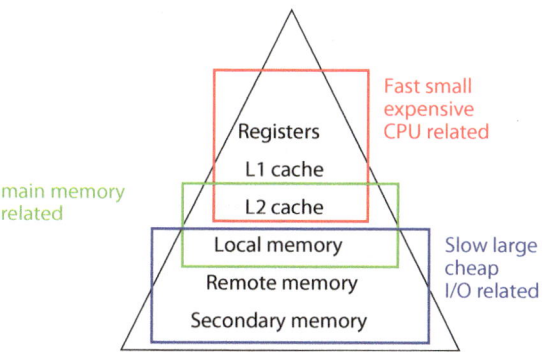

Fig. 3.1 The memory hierarchy

The *faster* (low latency) the memory, the more expensive and smaller it is. Typical latencies of local memories are 1 CPU cycle time (Level 1 cache with 8 kB), and 10 CPU cycle times for higher level caches of a few MB. The main memory RAM of a few GB is accessible in ca. 100 CPU cycle times. The local Hard Drive Disk (HDD) of up to 1 TB is a high latency memory of at least 10'000 CPU cycle periods, whereas cheap backup spaces (such as archiving types) have latencies of several seconds [s], but huge sizes of many TB. A hierarchical view of the different memory levels is shown in Figure 3.1.

Definition 8 *a* **communication device** *is a special hardware that allows a computational node to communicate with the other nodes through a private network or from one special node (often called the frontend) to an outside, public network. This hardware is often shortened by the acronym* **NIC** *(for Network Interface Controller).*

Networks can be characterized by their latencies, bandwidths, and by their topologies. We can distinguish between different levels of networking:

In SMP (Symmetric Memory Processors) machines the main memory access times are the same for all the processors, i.e. there is a low latency/high bandwidth memory switch through which all the processors are connected to the main memory. One talks about a symmetric or shared memory architecture.

In NUMA (Non-Uniform Memory Access) machines each compute element has its own local main memory, but the operating system considers the distributed memory architecture as to be virtually shared. Such machines need tightly coupled networks between the processors, otherwise the memory access times become very different between local and non-local memories.

In distributed memory architectures such as clusters, each node has its own main memory subsystem. The connection between the nodes is made by an internode communication network. This network can have different topologies (switches, fat trees, hypercubes, K-rings, tori, meshes, buses).

The connectivity between different resources on the same site are made through an internal network and the connections between sites are generally made through the internet or through a special purpose WAN.

Definition 9 *a* **System** *includes all the basic software needed to run a computer.*
A System is the Basic Software Level 1.

Definition 10 *a* **Thread** *is any executable in a core.*
A Thread is the Basic Software Level 2.

Definition 11 *an* **HPC application** *or simply an* **application** $a \in \mathcal{A}$ *is an executable that runs on a resource* r_i.
a is the Application Level 1.

Definition 12 *An application consists of k* **components** $a_k \in a$.
A Component is the Application Level 2.

Most of the current HPC applications are single component applications.

Definition 13 *Each component can be decomposed in* $a_{sd} \in a_k$ *tasks or subdomains running on one virtual processor.*
a_{sd} *is the Application Level 3.*

A component can further be parallelized with OpenMP or MPI.

Definition 14 *An* **availability** $\mathcal{A}_{k,j}(r_j, a_k, P_j, \tau^s_{k,j}, \tau^e_{k,j}), j = 1, \ldots, N_{avail}$ *is a set of* **holes in the input queues** *of the resources* r_j, **eligible** *to execute* a_k. *Here,* P_j *is the number of compute nodes available on* r_j *between the times* $\tau^s_{k,j}$ *and* $\tau^e_{k,j}$.

Definition 15 *A* **configuration** $\mathcal{C}_{k,j}(r_j, a_k, P_j, t^s_{k,j}, t^e_{k,j}, z), j = 1, \ldots, N_{conf}$ *is a set of* **best eligible resources** *to execute* a_k. *Here,* r_j *is the available resource,* a_k *is the application component,* P_j *the number of compute nodes reserved by* a_k, $t^s_{k,j}$ *is the start time (determined by the availability time* $\tau^s_{k,j}$*) and* $t^e_{k,j}$ *is the end time of execution predicted by a model or known. The difference* $t^e_{k,j} - t^s_{k,j}$ *is called* **execution time**. *The quantity z is the value of the cost function computed for an eligible resource.*

Definition 16 *a* **local policy** *is a set of rules* Rul_{r_i} *that are applied on a local resource* r_i *(not on the global Grid).*

The Grid is defined as $\mathcal{G}(s_\ell(r_i))$

$$\mathcal{G} = \cup_{\ell=1}^{N_{site}}(s_\ell) \tag{3.1}$$
$$s_\ell = \cup_{i=1}^{M}(r_i)$$

A community of users $\mathcal{U}$ is authorized to submit a set of applications a on the resources of the Grid $\mathcal{G}$.

Definition 17 *a* **Grid user** *or simply a* **user** $u \in \mathcal{U}$ *is an actor (entity) who has the rights to use at least one resource of the Grid* $\mathcal{G}$.

The security aspects are not tackled in this book. We assume that there is a high-level middleware that deals with the local security policies.

3.2 Parameterization of applications

3.2.1 Application parameter set

Let us suppose that an application has been cut into different components that weakly interact through a loosely coupled internode communication network. Then, these components can run quite independently on different computers interconnected by the internet or by a LAN in a university. Each component can be a parallel program.

It is supposed that a parallel application component a_k is well equilibrated, i.e. each parallel task takes close to the same execution and communication times. In this case, the component can be characterized by:

- N_j: Application input parameters requested to predict complexities
- O: Number of operations [GF]
- W: Number of words accessed from or written to main memory [Gw]
- Z: Number of messages sent over the communication network
- S: Number of words sent over the communication network [Gw]
- σ: Average message size [w]
- $V_a = O/W$ [F/w]
- $\gamma_a = O/S$ [F/w]
- n_{sd}: Number of parallel tasks (often subdomains) into which the component can be decomposed

In these definitions, the index k for a component has been omitted. In fact, in measurements, the quantities are mostly related to components, in optimization procedures the quantities are often related to algorithms, kernels, or simply to loops. We propose here to use "word" instead of "operand".

One *Flop* (F) is a 64bit (or 8 Bytes (B)) operation such as an add (ADD) or a multiply (MUL). A *Word* (w) consists of 64bits or 8 B. This implies that all operations are performed in double precision. We shall abbreviate by F (Flops), by w (Words), GF (1 *gigaFlop*=10^9 Flops), TF (1 *teraFlop* =10^{12} Flops), PF (1 *petaFlop* =10^{15} Flops), B (Bytes), kB (= 10^3 Bytes), MB/s (= 10^6 Bytes per second), W (Watts).

The average message size

$$\sigma = \frac{S}{Z}, \qquad (3.2)$$

is an important parameter to estimate the real memory bandwidth in a communication network. The parameter V_a [F/w] can be defined as [69]

$$V_a = \frac{O}{W}, \qquad (3.3)$$

and γ_a [F/w] by

$$\gamma_a = \frac{O}{S}. \tag{3.4}$$

These two parameters V_a and γ_a are critical to main memory access and network communication, respectively. In fact, a programmer should maximize V_a and γ_a. The bigger V_a the closer processor performance is at peak, whereas small values of V_a characterize components that are dominated by main memory bandwidth. The bigger γ_a, the less critical communication is. Since the latency can play an important role, Z should be as small as possible, and message size S is then as large as possible.

The quantity n_{sd} is often defined by a domain decomposition technique in which the geometry is divided into n_{sd} subdomains. This decomposition should be done such that each subdomain can be discretized by a structured mesh. Then, stride 1 operations can be activated, and best node performance efficiencies can be achieved.

3.2.2 Parameterization of BLAS library routines

In 1979, BLAS (**B**asic **L**inear **A**lgebra **S**ubprograms) library was proposed by Lawson et al. [93]. The idea behind was to simplify the usage of basic linear algebra operations. A classification of the BLAS routines were proposed: BLAS1 (vector-vector operations), BLAS2 (matrix-vector operations), and BLAS3 (matrix-matrix operations).

3.2.2.1 BLAS1: Parameterization of vector operations

Let us explain V_a, Eq. (3.3), by the DAXPY operation (BLAS1 operation) $\mathbf{y} = \mathbf{y} + a * \mathbf{x}$ (in double precision)[1], where $\mathbf{x}$ and $\mathbf{y}$ are considered to be long vectors with N components, stored sequentially in main memory, and the quantity a is a constant. The number of operations (MUL and ADD) is $O = 2N$, and the number of loads (load of vectors $\mathbf{x}$ and $\mathbf{y}$ from main memory) and stores (store $\mathbf{y}$ back to main memory) is $W = 3N$. If the vectors are not in cache, $V_a = 2/3$ and the expected peak performance is limited to two thirds of the peak main memory bandwidth in 64bit words per second. Note that the Cray Y-MP 8 processor SMP vector machine was specially designed for this operation with $V_\infty = 2/3$, Eq. (3.10).

3.2.2.2 BLAS2: Parameterization of matrix*vector operations

Slightly better values of V_a can be achieved with matrix times vector operations (BLAS2). If the multiplying matrix of rank N is full (i.e. a DGEMV, Double precision GEneral Matrix times Vector operation in LaPACK), the total number of

---

[1] all the BLAS operations are preceded by a 'S' in single precision by a 'D' in double precision: SAXPY-DAXPY, SGEMV-DGEMV SGEMM-DGEMM, etc..

operations is given by $O = 2N^2$ and the number of memory accesses by $W = N^2$, leading to $V_a = 2$. If the matrix is sparse, $V_a < 2$.

We shall see later that in multiple DO-loop constructions, and if the vector length is small enough, the vector can some times be kept in cache and reused during multiple DO-loop constructions. This increases the value of V_a and, accordingly, the performance of the algorithm.

3.2.2.3 BLAS3: Parameterization of matrix*matrix operations

A full matrix times full matrix operation, called DGEMM (Double precision GEneral Matrix times Matrix kernel), can be executed in the way shown in Fig. 3.2. The matrix A is subdivided in $M \times L$ blocks A_{IK} of size $m \times \ell$, B is subdivided in $L \times N$ blocks B_{KJ} of size $\ell \times n$, and the resulting matrix C is subdivided in $M \times N$ blocks C_{IJ} of size $m \times n$. The matrix operation

$$C = AB \tag{3.5}$$

can then be written as

$$C = \sum_{I=1}^{M} \sum_{J=1}^{N} C_{IJ} = \sum_{I=1}^{M} \sum_{J=1}^{N} \sum_{K=1}^{L} A_{IK} B_{KJ} . \tag{3.6}$$

The sizes ℓ, m, and n are chosen such that the blocks A_{IK}, B_{KJ}, and C_{IJ} can be kept in cache. The number of operations O_{block} (multiplies plus adds) and the number of main memory accesses W_{block} to calculate one C_{IJ} block are

$$O_{block} = 2nm\ell L \tag{3.7}$$
$$W_{block} = (n + m)\ell L ,$$

and the total number of operations and main memory accesses are:

$$O = 2nmNM\ell L \tag{3.8}$$
$$W = (n + m)NM\ell L .$$

Fig. 3.2 $C = A * B$: The block decomposition of a matrix * matrix multiplication module. The local block sizes can be kept in the highest level cache

This leads to

$$V_a = \frac{O}{W} = \frac{2nmNM\ell L}{(n+m)NM\ell L} = \frac{2nm}{n+m}. \tag{3.9}$$

Typical block sizes that can be kept in cache are $n = m = \ell = 100$, leading to $V_a = 100$. This big V_a value makes the DGEMM operation to be processor dominant. It is at the origin of the measured peak performances of the Linpack test case (called High Performance Linpack, HPL) used for the TOP500 list.

3.2.3 SMXV: Parameterization of sparse matrix*vector operation

When a Poisson equation is approximated by finite elements in a square domain discretized by a structured mesh in both directions, the resulting matrix has the form shown in Fig. 3.3. This matrix has three tri-diagonal bands with a distance corresponding to the number of unknowns in the first direction. In the same Figure, it is shown how the matrix can be compressed to a 9*N matrix, where N is the total number of unknowns. With an iterative solver, the matrix*vector operation is the most CPU time consuming part of an iteration step. One such sparse matrix*vector multiply demands altogether 18*N floating point operations (9*N multiplies and 9*N adds).

When combining the computation of a tri-diagonal block as a single operation, the number of operations per tri-diagonal block is 6*N (3*N adds and 3*N multiplies). The product of the tri-diagonal matrix*vector product is added to the old vector and the new vector is stored. Altogether there are 6*N operands to be loaded or stored, and the quantity $V_a = 1$, see Fig. 5.9. Thus, this operation is a very good candidate to measure the real main memory bandwidth. The performance in GF/s corresponds to the achievable main memory bandwidth in Gw/s.

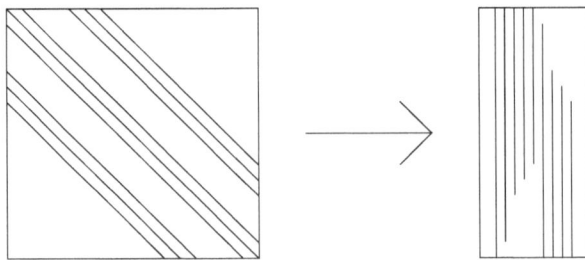

Fig. 3.3 Structure of the initial and compressed Poisson matrix

3.3 Parameterization of a computational nodes $P_i \in r_i$

A computational node P_i of resource r_i can be characterized by a number of parameters:.

- F_{core}: Core operating frequency [GHz]
- ω_{core}: Number of functional units
- N_{core}: Number of cores per processor
- N_{CPU}: Number of processors per node
- m_{node}: Main memory size per node in number of words [Gw]
- R_∞: Peak processor performance of a node [GF/s]
- R_∞^*: Peak processor performance of the attached unit in a node [GF/s]
- M_∞: Peak main memory bandwidth of the node [Gw/s]
- $V_\infty = R_\infty / M_\infty$ [F/w]
- J: Power of a node [W]

A computational node (Definition 4) consists of N_{CPU} processors (Definition 5), each has N_{core} cores (Definition 6), a memory subsystem (Definition 7) that can be shared or virtually shared, of devices for communication (Definition 8), and of a motherboard. The Northbridge of the motherboard connects the processor to the main memory and to the graphics card, and the Southbridge connects to the I/O subsystem and to the communication networks, see Fig. 3.4. Data related to the attached hardware (for instance SSE in Intel nodes) are marked with an upper * indication.

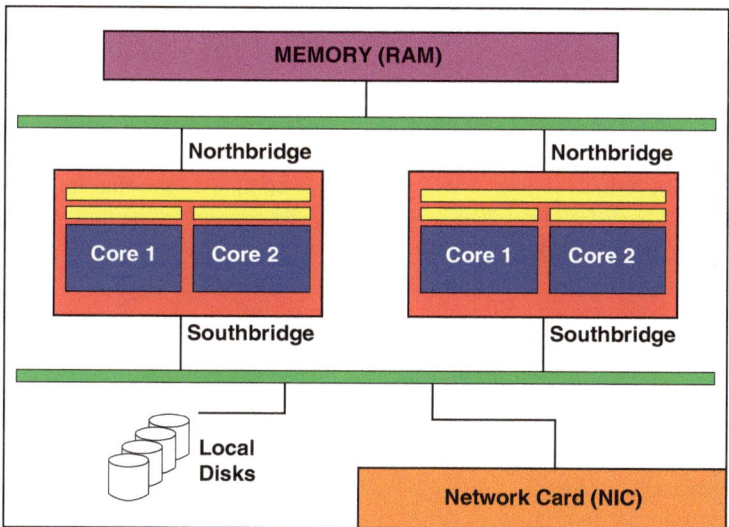

Fig. 3.4 A typical node. Two CPU's are shown (red). Each CPU has 2 cores (blue). Local cache memory is in yellow (L1 is local to the core, L2 is shared among the two cores). The two buses (north and south) are in green. The high level memory is in purple while the Network Interface Connection (NIC) is in orange. The local hard disk drives (HDD) are also depictured

A shared memory node with $N_{core} > 1$ and $N_{CPU} = 1$ is called an SMP (Symmetric Memory Processor) node. The main memory subsytem is common, and each core has the same access time. If $N_{CPU} > 1$ we talk about a NUMA (Non-Uniform Memory Access) node, the system accesses the distributed memory as if it is shared. This can lead to memory access slow down if data is not stored in the local memory of the executing processor.

The trend today goes towards *many-cores*. Chip makers produce processors with more and more cores. Processors on the market in 2008 have up to 8 cores (Cell BE processor), and 24 cores in 2010. GPUs will have up to 512 cores (NVidia CUDA Architecture codename "Fermi"). Application developers have to prepare to upcoming many-core architectures that could become quite complex. For instance, an IBM Cell node in the Sony Playstation3 has a main core with a main memory. The threads (code plus data) have to be prepared on the main core and distributed over a very fast bus among the eight computational cores with small local memories. Another example is the NVIDIA Tesla architecture with four GeForce GTX processors each one has 240 streaming-processor cores and an overall peak performance of 933 GF/s. Such particular architectures demand rewriting most of the HPC applications, for each architecture differently. How to use such complex and new architectures efficiently is not yet commonly known. First ideas can be found in chapter 7.

The value of

$$R_\infty = \omega_{core} N_{core} N_{CPU} \ F_{core}$$

can vary on today's multi-core architectures. Chip founders introduce a mechanism that automatically over-clocks one core for a single threaded application. Intel for instance applies IDA (Intel Dynamic Acceleration) on its new Core2Duo/Quad processors. For a 2.4 GHz dual core processor ($R_\infty = 4.8$ GF/s per core), IDA over-clocks one of the cores in case of a single threaded application to 2.6 GHz (thus $R_\infty = 5.2$ GF/s per core). This increases energy consumption. In many circumstances it would be more reasonable to adapt the node frequencies to the needs of the application and reduce energy consumption. We shall see later that under-clocking is possible and the core frequency can indeed be dynamically adapted to application needs (see section 8.5) without performance loss.

The quantity

$$V_\infty = \frac{R_\infty}{M_\infty}. \tag{3.10}$$

is the maximum number of operations a node can achieve during one **LOAD** or one **STORE**, i.e. during the time needed to move one operand from main memory to the lowest level cache or back.

On some nodes, two or more cores share a local memory (a level 2 or a level 3 memory cache). Table 3.1 shows some of the processors on the market in 2009.

The memory bandwidth per DDR memory module is 3.2 GB/s each, the two modules together can run up to 6.4 GB/s, corresponding to M_∞=0.8 Gw/s. The peak memory bandwidth M_∞ of a bank varies and is also subject to development by the chip founders. Table 3.2 shows the different memory characteristics.

Table 3.1 Major parameters for a selection of processors. Only main computational units are considered, not the special purpose attached units (such as SSE). Each processor exists in multiple models, those listed are used in the validation further. Note that the L1-cache value is the sum of all L1 caches of all cores. The L1 cache is in general divided by 2: half for the instructions the other for the data. ^{a} for the Itanium processor, there are 3 cache levels (L3, L2, L1) cache memories. ^{b} this is the processor of the Blue Gene/L machine ($\omega_{core} = 4$ if memory is aligned, using the *double hummer* technology, but $\omega_{core} = 2$ if the memory is not aligned)

Processor	am_{core}	N_{CPU}	N_{core}	ω_{core}	F [GHz]	R_∞ [GF/s]	R_∞^* [GF/s]	L2 [MB]	J [W]	M_∞ [Gw/s]	V_∞
Intel Pentium 4 HT	32	1	1	2	2.8	5.6		0.5	68.4	0.8	7
Intel Xeon 3000	64	1	1	2	2.8	5.6		1	103	0.8	7
AMD Opteron AM2	64	2	1	2	1.8	3.6	7.2	2	65	1.33	5.4
Intel Xeon 5150	64	1	2	2	2.8	5.6	11.2	4	65	1.33	8.4
Intel T7700	32	1	2	2	2.4	4.8	9.6	4	35	0.8	6
Intel Xeon 5310	64	1	4	2	2.7	10.8	21.6	4	180	8	1.35
Intel Itanium 9150	64	1	2	4	1.66	6.66		24^a	104	1.33	5
PowerPC 440^b	32	2	1	4	0.7	2.8		2	5	0.67	8.4

Table 3.2 SDRAM (Synchronous Dynamic Random Access Memory) types and characteristics to define M_∞

Standard Name	I/O Bus clock [MHz]	M_∞ [GB/s]
DDR-400	400	3.2
DDR2-800	400	6.4
DDR2-1066	533	8.53
DDR3-800	400	6.4
DDR3-1066	533	8.53
DDR3-1333	667	10.67
DDR3-1600	800	12.80

3.4 Parameterization of the interconnection networks

3.4.1 Types of networks

Two kinds of networks exist in our model: A **private network** for inter-nodes communication, and a **public network** for communication between resources. We note in addition to the Definition 8 that all the resources r_i of a Grid $\mathcal{G}$ site are connected to the Internet through a special communication device. In general, the resources r_i are accessible by a so-called *frontend machine*. A frontend machine is able to communicate with the outside (Internet) and through a private network to the computational nodes.

The IT literature commonly defines the different network types by acronyms that represent either their scale, their topology, their functionality, or even their protocol; all these definitions are from the OSI 7 layers model[82], for instance, LAN (for **L**ocal **A**rea **N**etwork) or WAN (for **W**ide **A**rea **N**etwork) for the scale of the network, P2P (for **P**eer-to-**P**eer), or client-server for their functionality, a TCP/IP network, or a proprietary Myrinet network for their protocol.

Table 3.3 Different interconnection network types (and device card) of several computational nodes on Grid resources. FE is FastEthernet, GbE is Gigabit Ethernet, My is Myrinet, IB is Infiniband, FT is a FatTree, 3D Torus is a network that connects the nodes in a torus manner, SB is a special network bit-to-bit that is dedicated to the MPI barrier treatment, NUMA is a NUMA-link network, and SS is for the Cray SeaStar 3D network. 'a' represents a network dedicated to administration, 'c' a computation network (for MPI messages for instance) and 'b' a special-purpose barrier network

Resource	Location	FE	GbE	My	IB	FT	3D Torus	SB	NUMA	SS (3D Torus)
Pleiades 1	EPFL	ac								
Pleiades 2	EPFL		ac							
Pleiades 2+	EPFL		ac							
Mizar cluster	EPFL		a	c						
Mizar SGI	EPFL		a						c	
Alcor	EPFL		a		c					
Blue Gene/L	EPFL		a			c	c	b		
Cray XT3	CSCS		a				c			c
IBM	CSCS		a			c			c	

Here, we parameterize the resources as a group of nodes interconnected by one or many networks taking care of the topology and the functionality of the network. For instance, we admit that an administration network is not used for computation.

Table 3.3 shows several machines and their communication networks that were accessible by researchers at EPFL mid 2008. The Pleiades machines were installed in the Institute of Mechanical Engineering, the other four machines at EPFL were installed at the computing centre, whereas the last two machines were installed at the Swiss National Supercomputing Centre at Manno.

Figure 3.5 shows a typical cluster with the different networks and clients accessing the cluster frontend for submitting jobs locally and remotely.

3.4.2 Parameterization of clusters and networks

In this book, we assume that a resource r_i is a homogeneous machine, for instance, a cluster made of nodes with the same characteristics, SMP machines, NUMA machines, or an individual machine.

A **Grid resource** (or simply **resource**) r_i is characterized by :

- P_i: Number of nodes connected to the network
- R_i: Total peak performance of the resource r_i [GF/s]
- m_i: Total memory of the resource r_i [Gw]
- C_∞: Maximum bandwidth of one link [Gw/s]
- ℓ_i: Total number of networks links
- b_i: Network bandwidth per node [Gw/s]
- $V_c = R_\infty/b_i$ [F/w]
- B_i: Message size such that transfer time = latency time [w]
- C_i: Total peak internode communication network communication bandwidth [Gw/s]

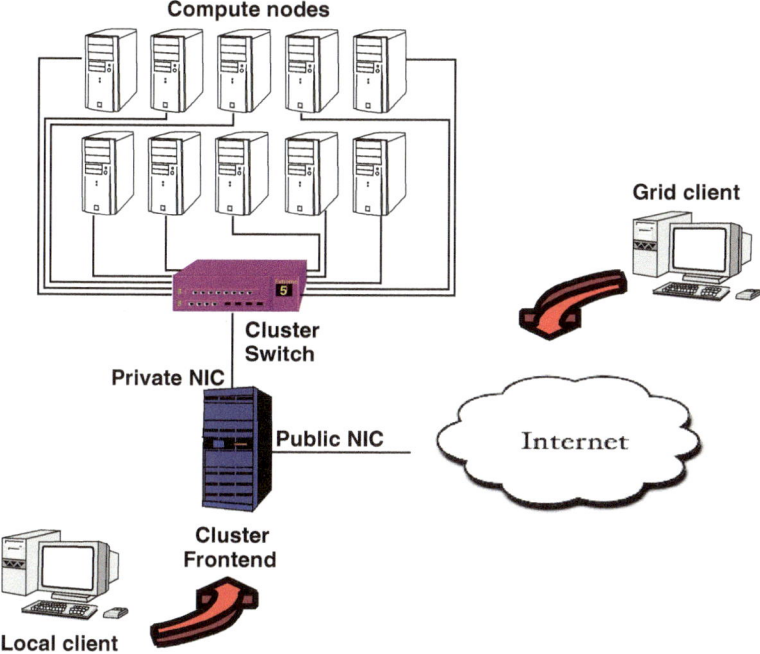

Fig. 3.5 A typical cluster configuration with a frontend (blue machine) and two clients : one accessing from the internal network, the other from the internet. Both uses the frontend machine through its public network interface

- L_i: Latency of the network [s]
- D_i: Maximum distance of two nodes in the interconnection network
- $\langle d \rangle_i$: Average distance
- H_i: Bisectional bandwidth

Then,

$$R_i = P_i R_\infty \tag{3.11}$$
$$m_i = P_i m_{node}$$
$$C_i = \ell_i C_\infty .$$

One can in addition define

$$b_i = \frac{C_i}{P_i \langle d \rangle_i} = \frac{\ell_i C_\infty}{P_i \langle d \rangle_i} \tag{3.12}$$

to be the average network communication bandwidth per node [Gw/s], by

$$V_c = \frac{R_\infty}{b_i} \tag{3.13}$$

the number of operations the processor can maximally perform during the time needed to send one operand from one node to another node [F/w], and by

$$B_i = b_i L_i \, . \tag{3.14}$$

This latter parameter B_i [w] is the message size that takes one latency time to be transferred from one node to another node. The average distance $\langle d \rangle_i$ measures the average number of links a message has to cross.

3.5 Parameters related to running applications

Different applications running on a node and/or need internode communications develop different characteristic parameters that are presented in this section. They are:

- r_a: Peak node processor performance for an application [GF/s]
- R_a: Real node processor performance for an application [GF/s]
- g_p: Efficiency of a core
- M_m: Real main memory bandwidth [GF/s]
- g_m: Efficiency of main memory
- $V_m = V_\infty \, / \, g_m$ [F/w]
- T: Total time for the execution [s]
- t_{comp}: CPU time [s]
- t_{comm}: Communication time [s]
- t_b: Data transfer time [s]
- t_L: Latency time [s]
- t_s: Start time of the execution [s]
- t_e: End time of the execution [s]
- γ_m: Number of operations possible during the time to transfer one operand through the network [F/w]
- Γ: Computing time over communication time
- E: Time fraction for computing
- A: Speedup, A in relation to Amdahl's law
- $P_{\frac{1}{2}}$: Number of nodes giving 50% of ideal speedup to a_k

The peak performance per node of an application component can be characterized by

$$r_a = \min(R_\infty, M_\infty V_a) = R_\infty \min\left(1, \frac{V_a}{V_\infty}\right) . \tag{3.15}$$

Here, the quantity r_a is called peak performance, i.e., the performance an application component can never reach.

The total measured main memory bandwidth M_m defines the main memory access related efficiency

$$g_m = \frac{M_m}{M_\infty} . \tag{3.16}$$

A good candidate to determine M_m is the SMXV kernel presented in section 3.2.3. SMXV is executed concurrently with as many substances as there are cores in the node. The measured total GF/s rate R_a of a node corresponds directly to the main memory bandwidth in Gw/s, i.e. $M_m = R_a$.

In NUMA machines, if main memory access is local, M_m is bigger than if data is attached to another processor. The parameter g_m also depends on the maturity of the motherboard and on the compiler. With the SMXV sparse matrix times vector program one measures $g_m = 0.8$ for an Itanium processor, $g_m = 0.5$ for a Pentium 4, and $g_m = 0.33$ for a Xeon 5150 (codename *Woodcrest*) node (see Tables 4.1 and 6.1 and Figure 3.4). In 2005, one substance of the SMXV benchmark was executed on a dual processor AMD Opteron node, $g_m = 0.48$ has been measured if memory is local, and $g_m = 0.37$ if the accessed data has to transit through the hyperchannel. If a NUMA architecture is used in a cluster, there is a big probability that one node runs at low performance, and for a well equilibrated parallel application the other nodes have to wait. As a consequence, the overall performance of a well equilibrated application is related to the node with lowest performance. In fact, when using the NUMA architecture of the old SGI Origin machines, NASA changed the memory allocation system to garantee that in an MPI application the main memory was always allocated locally, thus increasing efficiency of MPI applications by a factor of 3 in average, a factor measured on a machine running with the original operating system.

The quantity

$$V_m = V_\infty \, g_m \tag{3.17}$$

measures how much slower the main memory subsystem is with respect to the peak performance of the processors in a node. Typically, V_∞ is between 5 and 12 for RISC processors (see Table 3.1). However, if the memory access is not made in a sequential manner (i.e. with stride bigger than 1), the memory access time can be up to 10 times longer. Note that V_∞ is of the order of 1 or 2 for vector processors if there is no memory conflict. This is the major advantage of vector machines with respect to RISC nodes.

If

$$V_a > V_m , \tag{3.18}$$

the measured performance R_a defines the processor related efficiency

$$g_p = \frac{R_a}{R_\infty} \tag{3.19}$$

that is independent of M_∞ and N_{core}, but depends on the implementation of the algorithm. This quantity can vary quite substantially. For a Linpack benchmark that is dominated by matrix-matrix operations, $g_p > 1.7$, and $g_p^* > 0.85$ when running on the SSE hardware with $\omega_{core} = 4$ instead of $\omega_{core} = 2$ for the standard processor. Smaller g_p values are measured if this kernel is implemented by hand using three nested loops and then compile. With the most recent Intel compilers $g_p = 1.2$ ($g_p^* = 0.6$), i.e. they are able to translate specific code into SSE instructions when using the highest optimization levels, and the processor autodispatch code path '-axprocessor' ('-axSSE2' in the case of the MXM application kernel). If one tries to unroll the outermost loop by 4 the compiler can not further perform the SSE optimization, and $g_p = 0.4$. If the optimization option is switched off $g_p = 0.2$. In this case it is possible to help the compiler by unrolling the most outer loop, and $g_p = 0.4$.

We have to mention here that the estimations of g_m, Eq. (3.16), made with the SMXV benchmark program are only valid if the main memory is accessed in a pipelined manner, and with a stride equal to one. This means that accessed data must be stored contiguously in memory. Any jump in memory immediately drops the main memory bandwidth by up to one order of magnitude. This is due to the paging mechanism in x86 processors. This mechanism is different from the memory banking in vector machines where data can be accessed with any odd stride without reducing substantially main memory bandwidth.

Also per node, we can compute the CPU time t_{comp}, the time t_b for network communication, and the latency time t_L all in seconds [s] by

$$t_{comp} = \frac{O}{R_a} \tag{3.20}$$

$$t_b = \frac{S}{b_i}$$

$$t_L = L_i Z .$$

Since we suppose that the parallel tasks of the component are well balanced,

$$T = t_{comp} + t_b + t_L = t_{comp} + t_{comm} = t_e - t_s . \tag{3.21}$$

T corresponds to the time between the start of execution, t_s, and the end of execution, t_e, and t_{comm} is the total network communication time. Here, we suppose that no overlap between computation and communication is possible. This clearly overestimates the total execution time.

With

$$\gamma_m = \frac{R_a}{b_i} \left(1 + \frac{t_L}{t_b} \right) , \tag{3.22}$$

measuring the number of operations that can be performed during the time needed to send one operand over the network, we can introduce the quantity

$$\Gamma = \frac{\gamma_a}{\gamma_m} \tag{3.23}$$

that is at the origin of the so-called Γ-model [69]: Γ is 1 if the computation and the communication take the same time, $\Gamma = \infty$ if there is no communication. The quantity

$$E = \frac{t_c}{t_c + t_b + t_L}$$

measures the relation between communication to total time for an application component. Note that

$$\Gamma = \frac{E}{1 - E}$$

and the speedup

$$A = \frac{P}{1 + \frac{1}{\Gamma}} = PE \tag{3.24}$$

tells how much faster an application runs when P processors are used for the execution instead of one. $P_{\frac{1}{2}}$ corresponds to the number of nodes that gives half of the ideal scalability, i.e. for which $E = 0.5$, or $A = \frac{P}{2}$.

3.6 Conclusion

Many parameters defined in this chapter will be used to determine the complexities of the algorithms (chapter 4). Then, O, W, and V_a are important when optimizing loops for cores in chapter 5, and the cost function model in chapter 8 demands all the timing quantities and J.

Chapter 4
Models

"Man sollte die Dinge so einfach wie möglich machen, aber nicht noch einfacher."
Albert Einstein, German/Swiss/American Physicist

Abstract We present in this chapter a generic performance prediction model that has been validated for memory bound and CPU bound applications. Data during execution is monitored and reused to feed the model to predict execution and communication times.

4.1 The performance prediction model

To compute the execution time cost K_e, it is fundamental to predict how long the execution will last. Thus, it depends on the performance of a given application on a machine. It is supposed that the application-relevant parameters O, S, Z are machine-independent. They vary with the problem size. To express O, Z, and S as a function of input parameters that describe the job execution, complexity laws such as

$$O(N_1, N_2, N_3) = a_1 N_1 N_2^{a_2} N_3^{a_3} \tag{4.1}$$
$$Z(N_1, N_2, N_3) = b_1 N_1 N_2^{b_2} N_3^{b_3}$$
$$S(N_1, N_2, N_3) = c_1 N_1 N_2^{c_2} N_3^{c_3}$$

are proposed. The integer N_1 denotes for instance the number of time steps, the number of iterations, or, more generally, the number of times the algorithm is executed. The integers N_2 and N_3 describe the unknown complexity on parameters in an algorithm, for instance, the complexity with respect to a matrix size, or to the polynomial degree as in SpecuLOOS described later. These numbers must be specified by the user in the input files at job submission time.

The parameters a_i, b_i, c_i, $i = 1, \ldots, 3$ are computed by means of minimization processes using measurements made in the past. Measurements of O, Z, and S for at least three different values of N_i are needed to determine them. In this example we concentrate on the CPU time, thus finding a_i, $i = 1, 2, 3$. Since these parameters define the number of operations, they are independent of the hardware on which the execution has been made.

R. Gruber, V. Keller, *HPC@Green IT*,
DOI 10.1007/978-3-642-01789-6_4, © Springer-Verlag Berlin Heidelberg 2010

To find the complexities, Eqs. 4.1, the quantities O, S, and Z have to be monitored during execution. For instance for O, every Δt the number of operations performed is read by a Monitoring Module from a hardware counter existing in each core and stored in a database. At the end of the execution, these measurements are attributed to an application by interpreting the dayfiles. Suppose that there were N_1 time steps, the total number of operations is

$$O(N_1, N_2, N_3) = \sum_{j=1}^{N_1} O(j, N_2, N_3),\tag{4.2}$$

where $O(j, N_2, N_3)$ is the number of operations between time step $(j{-}1)\Delta t$ and $j\Delta t$ to be found in the database. The execution time $T_{exec} = N_1 \Delta t$ can be found at the end of the execution.

To improve the values of a_i and to adjust them to possible modifications in the hardware and the basic software, such as an improvement of the compiler or of the used libraries, these parameters are determined by an optimization procedure over N_{exec} runs for which execution data can be found on the database. The error function

$$\Phi = \sum_{l=1}^{N_{exec}} w_l (O_l - R_l P_l E_l T_l)^2\tag{4.3}$$

is minimized. Here, N_{exec} is the number of execution data that is considered, O_l is the number of operations for job number l, R_l is the average CPU performance (GF/s), P_l is the number of processors, E_l is the efficiency or the CPU usage, T_l is the measured CPU time used, and w_l is a weight. This weight is generally bigger for newer measurements. Triggering w_l makes it possible to also take into account changes in hardware, compilers, or libraries without reinitializing the model.

The minimization procedure consists of solving the nonlinear system of equations for a_1, a_2 and a_3:

$$\frac{\partial \Phi}{\partial a_1} = 0$$
$$\frac{\partial \Phi}{\partial a_2} = 0$$
$$\frac{\partial \Phi}{\partial a_3} = 0\tag{4.4}$$

The variable a_1 is linear and can be eliminated. The non-linear equations for a_2 and a_3 are solved using a Levenberg-Marquardt nonlinear least-squares algorithm [99, 101].

Generalization of the model

More generally, we can express the application related quantities (with any number of parameters) by:

$$O(N_i) = a_1 N_1 \prod_{i=2}^{\theta_O} N_i^{a_i} \tag{4.5}$$

$$Z(N_i) = b_1 N_1 \prod_{i=2}^{\theta_Z} N_i^{b_i}$$

$$S(N_i) = c_1 N_1 \prod_{i=2}^{\theta_S} N_i^{c_i} .$$

θ_O (resp. θ_Z and θ_S) is the number of major quantities that enter the complexity calculation. Thus, the minimization procedure consists of the resolution of the following system for the different parameters:

$$\frac{\partial \Phi}{\partial a_l} = 0, \ l = 1, \ldots, \theta_0 \tag{4.6}$$

where θ_0 is the minimum number of free parameters needed in the initialization process of the algorithm.

Let us present the complexity laws for the BLAS1, BLAS2, BLAS3 (see chapter 3.2.2), and the SMXV (see chapter 3.2.3) example. In all those cases, there is only one parameter that enters into the complexity law: for BLAS1, BLAS2, and SMXV it is the vector length N_2, and for the BLAS3 operation N_2 is the matrix size. In all cases, the two parameters $N_1 = N_3 = 1$.

BLAS1

The complexity laws of the number of operations O and the number of main memory accesses W of the SAXPY vector times vector operation, i.e. for $\mathbf{x} = \mathbf{y} + a * \mathbf{z}$, a being a constant, is

$$O(N_2) = 2 \cdot 10^{-9} N_2 \ [GF] \tag{4.7}$$
$$W(N_2) = 3 \cdot 10^{-9} N_2 \ [Gw] .$$

BLAS2

The complexity law for the DGEMV matrix times vector operation, i.e. for $A = B * \mathbf{x}$, A and B being full matrices, is

$$O(N_2) = 2 \cdot 10^{-9} N_2^2 \ [GF] \tag{4.8}$$
$$W(N_2) = 10^{-9} N_2^2 \ [Gw] .$$

Table 4.1 SMXV benchmark results for single precision arithmetic

Resource	R_∞ [GF/s]	M_∞ [Gw/s]	R_a for # Cores in GF/s				g_p	g_m
			1	2	4	8		
Pentium 4	5.6	1.6	0.8	-	-	-	0.14	0.5
Xeon	5.6	1.6	0.784	-	-	-	0.14	0.49
Woodcrest	21.3	2.66	1.397	1.687	1.743	-	0.08	0.33
Opteron	9.6	3.2	0.882	1.194	-	-	0.12	0.37
Itanium	5.2	1.6	1.31	1.29	-	-	0.25	0.81
Xeon 5300	37.3	8	0.936	1.594	1.736	1.744	0.05	0.22

BLAS3

The complexity law for the DGEMM matrix times matrix operation, i.e. for $A = B * C$, A, B, and C being full matrices, is

$$O(N_2) = 2 \cdot 10^{-9} \, N_2^3 \, [GF] \qquad (4.9)$$

$$W(N_2) = 10^{-9} \, \frac{n+m}{2 \cdot n \cdot m} \, N_2^3 \, [Gw] \,,$$

where n and m are the submatrix block sizes used in the highest level cache.

SMXV

The complexity laws for the number of operations and the number of main memory accesses for the sparse matrix times vector operation (see Fig. 3.3) are

$$O(N_2) = 18 \cdot 10^{-9} N_2 \, [GF] \qquad (4.10)$$

$$W(N_2) = 18 \cdot 10^{-9} N_2 \, [Gw] \,.$$

One realises that $O = W$, and, as a consequence, $r_a = M_\infty$. Thus, by measuring the processor performance R_a of the SMXV operation on a node one can get the quantity g_m measuring the main memory access efficiency

$$g_m = R_a / M_\infty \,. \qquad (4.11)$$

Typical measurements have been made on different nodes presented in Table 4.1. One observes that the efficiency $g_p = R_a / R_\infty$ decreases with an increasing number of cores. The same happens for g_m.

In order to apply the model, the Monitoring Module (MM) must be able to monitor the quantities shown in Table 4.2.

Table 4.2 Quantities that enter the performance prediction model and that must be monitored and stored in a System Information database

Quantity	Description	Who	Style
N_i	Applications' input parameters	user	manually
a_i, b_i, c_i	Prediction Model parameters	system	automatically
Z	Average number of MPI messages	system	automatically
S	Average size of MPI messages	system	automatically
O	Number of FLOPS	system	automatically
T_{exec}	Execution time	system	automatically
P_i	Number of compute nodes	system	automatically
E_i	Average efficiency of the nodes	system	automatically

4.2 The execution time evaluation model (ETEM)

When the complexities $O(N_i)$, $Z(N_i)$, and $S(N_i)$ are found, we can tackle the problem of estimating the CPU time for a case with different values of the N_i parameters by simply applying the formula in Eq. (4.1). With the parameters and measurements $t_c^{(1)}$, $R_\infty^{(1)}$, $M_\infty^{(1)}$, $R_a^{(1)}$, $g_m^{(1)}$, and $P^{(1)}$ on a known resource r_1 where the application has been executed previously, it is possible to estimate the CPU time on a second resource r_2 on which the application has never run. For an application component dominated by the memory subsystem, the real performance, the computation and communication times, and the total time can be estimated on the second machine by

$$R_a^{(2)} = R_a^{(1)} \frac{g_m^{(2)} M_\infty^{(2)}}{g_m^{(1)} M_\infty^{(1)}}$$

$$t_{comp}^{(2)} = \frac{O}{R_a^{(2)} P} \qquad (4.12)$$

$$t_{comm}^{(2)} = \frac{S}{b_a^{(2)}(S, Z)} + LZ$$

$$T_{exec}^{(2)} = t_{comp}^{(2)} + t_{comm}^{(2)} .$$

If the algorithm is dominated by processor performance and not by main memory bandwidth, i.e. when $g_m V_a > g_p V_m$, then, in Eq. (4.12), M_∞ has to be replaced by R_∞, and g_m by g_p.

4.3 A network performance model

The quantity γ_m has been defined using the maximum interconnection bandwidth b_∞ and the latency L, or the relative network bandwidth (which includes the communication time due to latency) b_a, Eq. (3.22). Let us present a model that is able to predict b_a in function of the average size of the sent messages.

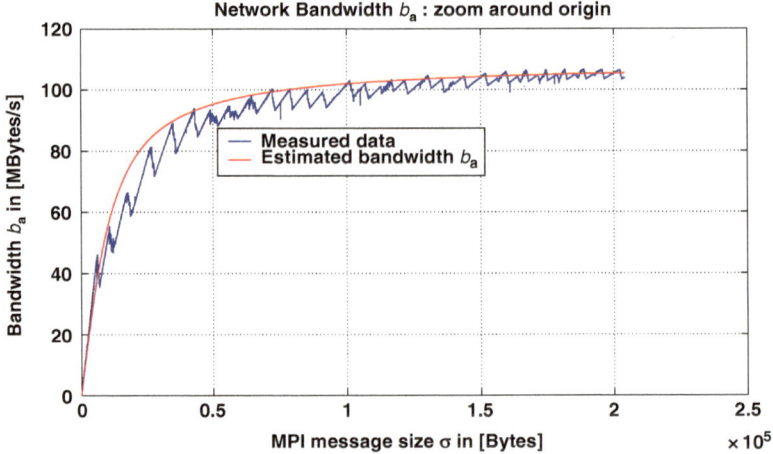

Fig. 4.1 The bandwidth b_a of the GbE network as a function of the size of the MPI messages σ. The blue curve shows the measured values, the red curve is the estimation made from Eq. (4.13) with $x = 69.26$ and $y = 1.01 \cdot 10^{-4}$. The implementation used is MPICH2 version 1.0.5 compiled with the Intel C++ *icpc* compiler

The average communication bandwidth b_a for an application can be estimated when the average message size σ, Eq. (3.2) is known. Using the $b_a(\sigma)$ curve, those presented in Fig. 4.1 for the GbE shows that b_a strongly depends on the message size. Comparing with the $b_a(\sigma)$ curve for a Myrinet interconnect, Fig. 4.2 shows that such curves can much differ from one network to another one. In our model, the $b_a(\sigma)$ curve can approximated by

$$b_a = x\, arctan(y\sigma). \tag{4.13}$$

The latency time is given by ($\sigma = 1$)

$$L = \frac{1}{x\, arctan(y)} \tag{4.14}$$

and the maximal bandwith ($\sigma = \infty$)

$$b_{max} = \frac{\pi}{2}\, x. \tag{4.15}$$

The values of x, y are obtained by 2 ping-pong benchmark measurements performed for each network with messages of sizes 1 and 10^9 Bytes. We observe that it is mandatory to weight more the very small message size and the very large one. In between, the curve approximates well the measurements. Figure 4.1 shows the model applied for a GbE network while Figure 4.2 represents the Myrinet network estimation. The values for the constants x, y are given in Table 4.3.

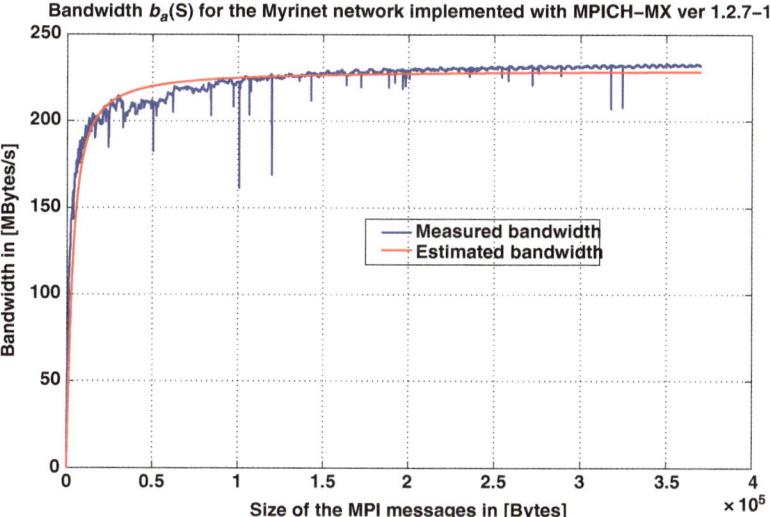

Fig. 4.2 The bandwidth of the Myrinet network as a function of the size of the MPI messages. The blue curve shows the measured values, the red curve is the estimation made from Eq. (4.13) with $x = 146.36$ and $y = 2.92 \cdot 10^{-04}$. The implementation used is MPICH-GM version 1.2.7-1 compiled with the GNU g++ compiler

Most often, another model to determine

$$b_a = \frac{1}{L + \frac{\sigma}{b_{max}}} \tag{4.16}$$

is used.

Both models represent well the measured curves (see Table 4.3).

Table 4.3 Values of the constants x and y from Eq. (4.13) for different interconnection networks

Network	Implementation	x [MB/s]	y [s]
GbE	MPICH2 (ver 1.0.5)	69.26	$1.01 \cdot 10^{-4}$
FE	MPICH1 (ver 1.2.5)	6.71	$2.51 \cdot 10^{-4}$
Myrinet	MPICH-GM (ver 1.2.7-1)	146.36	$2.92 \cdot 10^{-4}$
Blue Gene/L	MPICH	141.26	$6.33 \cdot 10^{-4}$

4.4 The extended $\Gamma - \kappa$ model

It is now possible to present the $\Gamma - \kappa$ model. This model extends the capabilities of the Γ model where:

$$\Gamma = \frac{\gamma_a}{\gamma_m} . \tag{4.17}$$

Table 4.4 Decision table for the application's fitness based on their needs. Best suited networks are for $\kappa > 1$ and $\Gamma > 1$

	$\Gamma \leq 1$	$\Gamma > 1$
$\kappa \leq 1$	The node performance is bounded by the peak memory bandwidth $M_\infty \cdot t_{comm} \geq t_{comp}$.	The node performance is bounded by the peak memory bandwidth $M_\infty \cdot t_{comm} \ll t_{comp}$
$\kappa > 1$	The node performance is bounded by the peak CPU performance $R_\infty \cdot t_{comm} \geq t_{comp}$.	The node performance is bounded by the peak CPU performance $R_\infty \cdot t_{comm} \ll t_{comp}$.

We remind that the quantity γ_a expresses the *network communication needs of the application* and γ_m defines *what the network can give*. The Γ value indicates if a network is sufficiently performing for the application. In fact, if $\Gamma = 1$, CPU and communication times are equal, if $\Gamma = 0$, communication takes all the time, if $\Gamma = \infty$, there is no internode communication.

We also define the alternative quantity

$$\kappa = \frac{V_a}{V_m}.$$ (4.18)

The quantity V_a expresses the *main memory bandwidth needs of the application* and V_m defines the *maximum main memory bandwidth of a node*.

We claim that using Eqs. (4.17)-(4.18), it is possible to express the suitability of the resources to the application needs. Table 4.4 summarizes the model. One can easily see that given Γ, it is possible to determine if the application is parallel or not (or embarrassingly parallel); given κ, we can determine whether the application needs a high memory bandwidth or a high peak performance processor node.

4.5 Validation of the models

In Section 4.1 page 49 we present a model which predicts the scaling rules of a given application. The results from the three testbed applications are presented here.

4.5.1 Methodology

For each application, a set of test-cases is chosen. The application is then run for each test-case on a specific reference machine. We chose Pleiades2 as the reference machine. We find the scaling laws for O_k, S_k, and Z_k in the case of a parallel application. We then estimate the values on the other resources in the testbed knowing these quantities. Finally, the execution time T_{exec} is predicted. All the test-cases are then run on each resource, the results are compared with the estimated values. The computation of the error is as shown in Eq. (4.19). The error between the estimated quantity Q_e and the measured one Q_m is

$$\delta = \frac{|(Q_m - Q_e)|}{Q_e}.$$ (4.19)

4.5.2 Example: The full matrix*matrix multiplication DGEMM

Table 4.5 shows the results for the DGEMM testbed application measured on one node of the Pleiades2 cluster. The complexity law for the number of operations O_k (in GF) is

$$O_k(N_1, N_2) = 2.00 \cdot 10^{-9} N_1 N_2^{3.00}$$

and perfectly fits the exact one, Eq. (4.9).

Table 4.5 Performance Prediction validation for the DGEMM application. The used machine was Pleiades2. The compiler used is Intel $ifort$ with the $-O3 - axW$ compilation flags and the Intel's optimized library MKL version 10.0 (DGEMM implementation). $g_p = 0.83$

| N_1 | N_2 | T_{exec} [s] | | | R_a [GF/s] | O_k [GF] | | | |
		meas.	estim.	δ	meas.	meas.	estim.	δ	g_p
4	1000	1.73	1.72	0.006	4.604	8.00	8.00	0.000	0.82
10	1000	4.31	4.30	0.002	4.631	19.99	20.00	0.001	0.83
15	1000	6.45	6.45	0.000	4.647	30.00	30.00	0.000	0.83
4	1500	5.92	5.81	0.019	4.557	26.99	27.00	0.000	0.82
10	1500	14.22	14.52	0.021	4.745	67.50	67.50	0.000	0.84
15	1500	21.33	21.78	0.018	4.746	101.24	101.25	0.000	0.84
4	2000	13.49	13.77	0.020	4.743	63.99	64.00	0.000	0.84
10	2000	33.47	34.42	0.028	4.779	159.99	160.00	0.000	0.85
15	2000	50.28	51.64	0.026	4.772	239.97	240.00	0.000	0.85
20	800	4.47	4.41	0.014	4.585	20.48	20.48	0.000	0.82
4	4000	106.49	110.15	0.033	4.808	511.99	512.00	0.000	0.86

The g_p value is computed using Eq. (3.19). It shows the accuracy of the implementation with the peak performance R_∞. The efficiency of the core g_p is shown to be more than 80 %. DGEMM is CPU performance dominated, not by memory bandwidth.

With the complexity laws, the correspondence between machines, Eq. (4.12) and the values of the considered processors, we are able to compute the performance R_a and the execution time T_{exec} on the other machines of the testbed. We show results on the Pleiades clusters where for Pleiades1: $g_p = 0.62$, on Pleiades2: $g_p = 0.83$ and on Pleiades2+: $g_p = 0.69$. Table 4.6 presents the predicted and measured values for the Pleiades1 resource while Table 4.7 are the values for the Pleiades2+ resource.

Table 4.6 DGEMM: measured on the Pleiades1 (meas.) and estimated (*estim.*) using results on Pleiades2. $g_p = 0.62$

N_1	N_2	T_{exec} [s] meas.	estim.	δ	R_a [GF/s] meas.	estim.	δ	O_k [GF] meas.	estim.	δ	g_p
4	1000	2.33	2.30	0.013	3.429	3.472	0.013	7.99	8.00	0.001	0.62
10	1000	5.83	5.76	0.012	3.428	3.472	0.013	20.00	20.00	0.000	0.61
15	1000	8.72	8.64	0.009	3.437	3.472	0.021	30.00	30.00	0.000	0.62
4	1500	7.83	7.78	0.019	3.446	3.472	0.010	26.99	27.00	0.000	0.62
10	1500	19.64	19.44	0.006	3.436	3.472	0.021	67.50	67.50	0.000	0.62
15	1500	29.41	29.16	0.009	3.442	3.472	0.010	101.24	101.25	0.000	0.62
4	2000	18.40	18.43	0.010	3.479	3.472	0.009	64.03	64.00	0.001	0.62
10	2000	45.94	46.08	0.002	3.483	3.472	0.002	160.02	160.00	0.000	0.62
15	2000	69.02	69.12	0.001	3.477	3.472	0.001	240.01	240.00	0.000	0.62
20	800	6.02	5.90	0.020	3.400	3.472	0.021	20.47	20.48	0.000	0.60
4	4000	147.05	147.47	0.003	3.482	3.472	0.003	512.04	512.00	0.000	0.62

Table 4.7 DGEMM using SSE library: measured on the Pleiades2+ (*meas.*) and estimated (*estim.*) using results on Pleiades2. $g_p^* = 0.69$, $R_\infty^* = 41.6$ GF/s

N_1	N_2	T_{exec} [s] meas.	estim.	δ	R_a^* [GF/s] meas.	estim.	δ	O [GF] meas.	estim.	δ	g_p^*
4	1000	0.32	0.27	0.185	24.81	29.4	0.156	7.99	8.00	0.001	0.58
10	1000	0.74	0.68	0.074	26.76	29.4	0.090	19.99	20.00	0.001	0.63
15	1000	1.14	1.02	0.118	26.15	29.4	0.120	30.00	30.00	0.000	0.61
4	1500	0.92	0.92	0.000	29.10	29.4	0.111	27.00	27.00	0.000	0.68
10	1500	2.32	2.30	0.009	28.98	29.4	0.014	67.50	67.50	0.000	0.68
15	1500	3.44	3.44	0.000	29.35	29.4	0.002	101.23	101.25	0.000	0.68
4	2000	2.00	2.18	0.083	31.90	29.4	0.085	64.01	64.00	0.000	0.75
10	2000	5.00	5.44	0.081	31.97	29.4	0.085	160.00	160.00	0.000	0.75
15	2000	7.49	8.16	0.082	32.02	29.4	0.089	240.01	240.00	0.000	0.75
20	800	0.77	0.69	0.116	26.37	29.4	0.103	20.47	20.48	0.001	0.62
4	4000	14.73	17.41	0.154	34.74	29.4	0.182	511.99	512.00	0.000	0.82

Figure 4.3 shows the variation of performance as a function of N_2 for a multi-cores based node of the Pleiades2+ cluster and for a single core based node of the Pleiades2 cluster. On Pleiades2, the peak performance around 4.5 GF/s is achieved with matrices of size $N_2 > 100$. On the Pleiades2+, the same observation is made when running with one thread, but approximatively 9 GF/s is achieved. For more than one thread, the matrix sizes must be much bigger when peak performance should be approached.

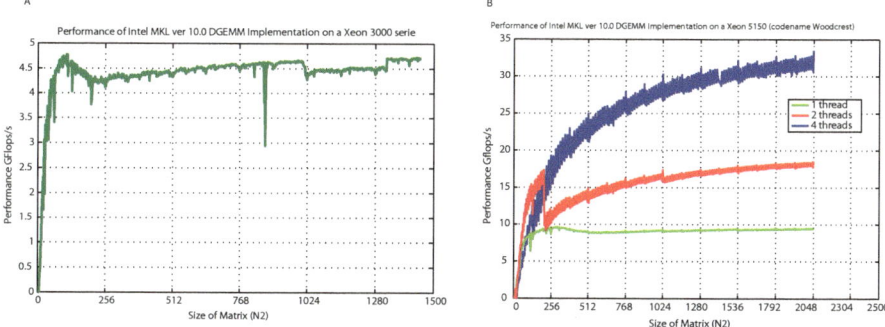

Fig. 4.3 Scaling performance of the Intel DGEMM implementation (MKL library version 10.0) in function of the size N_2 of the considered matrix. Left: performance on a node of Pleiades2, right on a node of Pleiades2+ with a variation of the number of running threads

4.5.3 Example: Sparse matrix*vector multiplication SMXV

Table 4.8 presents the results of SMXV on Pleiades2. With these values it is possible to apply the complexity law and compute the quantities g_m and r_a for the SMXV application.

Table 4.8 SMXV on Pleiades2: Performance Prediction validation for the SMXV application's building block using double precision data. The implementation of SMXV is written in Fortran. The compiler used is Intel $ifort$ without any optimization flags. $r_a = M_\infty V_a = 0.8*1 = 0.8\,GF/s$, $g_m = \frac{R_a}{r_a} = \frac{R_a}{0.8}$. Thus $g_m = 0.46$ and $R_a = 0.368\,GF/s$

| N_1 | N_2 | T_{exec} [s] | | | R_a [GF/s] | O_k [GF] | | | |
		meas.	estim.	δ	meas.	meas.	estim.	δ	g_m
10	$4 \cdot 10^6$	1.93	1.95	0.010	0.373	0.714	0.719	0.007	0.46
25	$4 \cdot 10^6$	4.81	4.88	0.014	0.375	1.798	1.798	0.000	0.46
100	$4 \cdot 10^6$	19.15	19.54	0.020	0.375	7.191	7.192	0.000	0.46
10	$16 \cdot 10^6$	8.16	7.82	0.133	0.354	2.871	2.879	0.003	0.44
25	$16 \cdot 10^6$	20.31	19.53	0.040	0.354	7.182	7.196	0.002	0.44
100	$16 \cdot 10^6$	81.06	78.14	0.037	0.355	28.758	28.786	0.009	0.44
10	$36 \cdot 10^6$	17.84	17.61	0.013	0.363	6.519	6.477	0.007	0.46
25	$36 \cdot 10^6$	44.10	44.00	0.002	0.368	16.209	16.194	0.000	0.46
100	$36 \cdot 10^6$	176.17	176.00	0.001	0.367	64.761	64.778	0.000	0.46

With the complexity law, Eq. (4.10) for the number of operations O_k it is possible to compute the number of operations for all the cases (N_1, N_2). With Eq. (3.16) it is possible to compute g_m on each resource, thus R_a for SMXV (using Eqs. (3.16), (4.12) and assuming that the target resource is unknown). Table 4.9 presents the results of SMXV on the Pleiades2+ resource while Table 4.10 presents SMXV on the Pleiades1 resource.

Table 4.9 SMXV: measured on one core of Pleiades2+ (*meas.*) and estimated (*estim.*) using results on Pleiades2. $g_m = 0.34$

N_1	N_2	T_{exec} [s]			R_a [GF/s]			O_k [GF]			g_m
		meas.	estim.	δ	meas.	estim.	δ	meas.	estim.	δ	
10	$4 \cdot 10^6$	1.54	1.55	0.006	0.464	0.464	0.000	0.715	0.719	0.006	0.34
25	$4 \cdot 10^6$	3.86	3.87	0.000	0.466	0.464	0.000	1.799	1.798	0.001	0.34
100	$4 \cdot 10^6$	15.40	15.50	0.006	0.467	0.464	0.006	7.192	7.192	0.000	0.34
10	$16 \cdot 10^6$	6.48	6.20	0.045	0.443	0.464	0.046	2.871	2.879	0.002	0.34
25	$16 \cdot 10^6$	16.14	15.49	0.023	0.445	0.464	0.041	7.182	7.196	0.001	0.34
100	$16 \cdot 10^6$	64.48	61.97	0.008	0.446	0.464	0.039	28.759	28.786	0.000	0.34
10	$36 \cdot 10^6$	13.99	13.96	0.002	0.466	0.464	0.004	6.519	6.477	0.007	0.34
25	$36 \cdot 10^6$	34.86	34.89	0.001	0.465	0.464	0.002	16.209	16.194	0.001	0.34
100	$36 \cdot 10^6$	139.27	139.58	0.002	0.465	0.464	0.002	64.761	64.778	0.000	0.34

Table 4.10 SMXV: measured on the Pleiades1 (*meas.*) and estimated (*estim.*) using results on Pleiades2. $g_m = 0.44$

N_1	N_2	T_{exec} [s]			R_a [GF/s]			O_k [GF]			g_m
		meas.	estim.	δ	meas.	estim.	δ	meas.	estim.	δ	
10	$4 \cdot 10^6$	1.93	1.97	0.020	0.371	0.365	0.016	0.716	0.719	0.004	0.44
25	$4 \cdot 10^6$	4.74	4.92	0.037	0.379	0.365	0.033	1.796	1.798	0.001	0.44
100	$4 \cdot 10^6$	19.28	19.70	0.021	0.373	0.365	0.022	7.191	7.192	0.000	0.44
10	$16 \cdot 10^6$	7.79	7.88	0.011	0.369	0.365	0.011	2.875	2.879	0.000	0.44
25	$16 \cdot 10^6$	20.52	19.70	0.042	0.351	0.365	0.038	7.203	7.196	0.000	0.44
100	$16 \cdot 10^6$	81.83	78.78	0.039	0.352	0.365	0.036	28.804	28.786	0.003	0.44
10	$36 \cdot 10^6$	-	17.75	-	-	0.365	-	-	6.477	-	0.44
25	$36 \cdot 10^6$	-	44.36	-	-	0.365	-	-	16.194	-	0.44
100	$36 \cdot 10^6$	-	177.45	-	-	0.365	-	-	64.778	-	0.44

Chapter 5
Core optimization

"Multicore puts us on the wrong side of the memory wall. Will chip-level multiprocessing ultimately be asphyxiated by the memory wall?"
Thomas Sterling, Father of the (modern) Beowulf

Abstract This chapter covers the techniques, skills and hints that are mandatory to optimize an application at the single-core level. Some useful notions such as memory hierarchy, data representation, floating point operations, pipelining and core architecture are described. The optimization techniques are then presented, such as unrolling, indirect adressing, cache-miss, dependencies and inlining. All the techniques are described in detail and emphasized with examples, even though compilers are doing this work automatically. Finally, optimization steps in thermonuclear fusion codes are discussed.

5.1 Some useful notions

5.1.1 Data hierarchy

Data in a computer is organized in a hierarchical manner (see Fig. 3.1 and Table 5.1). The registers run at processor speed, the caches, if there are, have a bandwidth of typically two 64 bit words (one load and one store) per cycle period (cp), and the latency time corresponds to a few cycle periods, higher level caches having higher latency (see Fig. 5.6). The main memory (DDR in Fig. 5.6) of a few GB is connected to the cache through the motherboard. The latency/bandwidth of the main memory is about one order of magnitude bigger/smaller than the ones of the caches.

A disk can store one to two orders of magnitudes more data than main local memories. The access time, or latency of a disk is given by the rotation frequency that varies between 5'600 and 20'000 turns/minute. This gives half turn times of 1.5 to 5 ms, corresponding to about 10^7 cycle periods of the processor. This indicates that disk access should be done rarely. The bandwidth is two orders of magnitude smaller than the main memory bandwidth.

The next level of data storage is given by systems that can satisfy storage requirements for hundred TB and more. The data is stored on cartridges or CDs that have to be loaded on a read/write station by a robot system. This takes about 30 seconds,

R. Gruber, V. Keller, *HPC@Green IT,*
DOI 10.1007/978-3-642-01789-6_5, © Springer-Verlag Berlin Heidelberg 2010

Table 5.1 Data close to the processor registers is fast but small, data on secondary memory such as disks or tapes is large but slow

Type of memory	Typical latency in cp	Typical bandwidth in words/cp	Typical size in words
Registers	1	2	2^3
L1 cache	2	2	2^{10}
L2 cache	2^2	2	2^{17}
Local memory	2^4	2^{-3}	2^{27}
Disk	2^{20}	2^{-7}	2^{34}
Tape	2^{30}	2^{-12}	2^{35}

thus, the latency corresponds to 10^{11} cycle periods. The bandwidth of such a device is comparable to the one for a disk. One cartridge can store up to 1 TB of data, cabinets can include thousands of cartridges. Such data units are used to archive data that has to be kept for a long time.

5.1.2 Data representation

Different types of data can be found in a computer. There are character strings in units of bytes, instructions with different representations on different machines, logical variables represented by bits and stored as bytes, integer numbers that can be 32 bits long, single precision floating point numbers that are 32 bits long, and double precision floating point numbers that are 64 bits long.

5.1.2.1 Big endian versus little endian

The data representation can be based on a little endian or a big endian standard [126]. In fact, this strange denomination (see [34]) comes from "Gulliver's Travels" story in which politicians made war about a dispute on from which side an egg should be broken, from the little or from the big end. Data is represented either from left to right (big endian) or from right to left (little endian). This can cause troubles if one writes data in binary form by a machine of one data type, and then reads this binary data by a machine of the other data type. One way to circumvent this problem is to write and read data in formatted form that is a unique standard. A double precision word, Fig. 5.1, takes 8 bytes when represented in binary form. In formatted form, the signs of the mantissa and of the exponent are one byte long, the absolute value of the exponent is represented by 4 bytes, and the mantissa takes

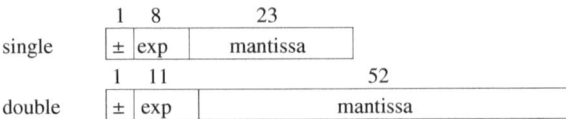

Fig. 5.1 The IEEE floating point standard 754-1985 (newly revised as IEEE 754-2008 standard) for single precision 32 bit floating point representation and double precision 64 bit numbers

16 bytes. This leads to 22 bytes instead of 8. Formatted writing takes almost three times as much storage than binary writing, and needs transformation from binary to formatted form and back. Another solution is to use the library routines that perform data transformation from one binary representation to the other. Those exist on all computers.

5.1.2.2 Address space addressing

Another discussion is going on for decades on the number of bits used to address the main memory. Machines with 32 bits addressing like the Intel Pentium 4 or the AMD Athlon 32 bits. Since the sign bit cannot be used by C++ and Fortran compilers, the directly addressable main memory space is limited to $2^{31} = 2$ GB. This limit has been left behind by all PC and server manufacturers. They now talk about 64 bit addressing, but in reality, only 40 bits (Athlon Opteron), or 44 bits (HP Alpha), or 48 bits (Intel Xeon) are used for addressing. In fact, machines with 40 bits addressing space can have main memories of up to half a TB. If we would like to address the entire main memory of a petaflop cluster as to be virtually shared, addressing would need at least 51 bits. In fact, 64 bits would make it possible to access 8 EB (exabytes) of data. Petaflop clusters already exist, exaflop computers should be ready by 2018. New virtually shared programming paradigms are under development. Computer manufacturers should start to be prepared for near future and extend address space to 64 bits.

5.1.2.3 Single versus double precision

Some processors perform all arithmetic operations in 64 bits, others produce two times more single precision results than double precision ones. Single precision words take half storage and half transfer time than double precision words. A single precision operand is often transformed into a double precision word before execution by the processor, and the result transformed back before storage. There are IEEE standards for the representation of the floating point numbers, one standard for a single precision floating point number using 32 bits and one standard for a double precision floating point number using 64 bits, Fig. 5.1. They are valid for big or little endian machines. Single precision exponents go from 2^{-126} to 2^{+128}, and the mantissa has 23 bits, i.e. 7 decimals. Double precision floating point numbers have exponents going from 2^{-1022} to 2^{+1024}, and the mantissa has 52 bits, i.e. 15 decimals.

By using double precision numbers one is on the safer side, but one has to access the double amount of data. For main memory access dominated applications, this increases the CPU time by up to a factor of two. Wrong results due to truncation errors then rarely occur. An example is demonstrated by the small matrix, Fig. 5.2. When computing the eigenvalues of this 4×4 matrix in single precision, the Lapack kernels SSYTRD and SPTEQR on a Pentium 4 predict one negative eigenvalue -2.131×10^{-5}. The matrix A seems not to be positive definite. If we run in double

$$A = \begin{pmatrix} 238 & -74 & 57 & -68 \\ -74 & 132 & -8 & -59 \\ 57 & -8 & 190 & -176 \\ -68 & -59 & -176 & 211 \end{pmatrix}$$

Fig. 5.2 Matrix A with an eigenvalue of 9.686×10^{-8}

precision, i.e. executing DSYTRD and DPTEQR on the same machine, all eigenvalues of A are positive. In fact, matrix A has been constructed starting with the positive quadratic form

$$
\begin{aligned}
2E = \; &(7x_1 - 11x_2 - 2x_3 + 7x_4)^2 \\
+ \; &(8x_1 + x_2 + 7x_3 - 9x_4)^2 \\
+ \; &(5x_1 - 3x_2 + 11x_3 - 9x_4)^2 \\
+ \; &(10x_1 + x_2 - 4x_3)^2
\end{aligned}
\tag{5.1}
$$

and looking for its minimum

$$
\begin{aligned}
\frac{\partial E}{\partial x_1} &= 238x_1 - 74x_2 + 57x_3 - 68x_4 \\
\frac{\partial E}{\partial x_2} &= -74x_1 + 132x_2 - 8x_3 - 59x_4 \\
\frac{\partial E}{\partial x_3} &= 57x_1 - 8x_2 + 190x_3 - 176x_4 \\
\frac{\partial E}{\partial x_4} &= -68x_1 - 59x_2 - 176x_3 + 211x_4 \, .
\end{aligned}
\tag{5.2}
$$

Since the 4 quadratic terms are not linearly dependent, A is positive definite. By the way, this ill-conditioned matrix has been constructed in the following way:

1. Choose four numbers $n_1{=}1493$, $n_2{=}421$, $n_3{=}108$, $n_4{=}383$.
   Then, $n_1 * n_4 - n_2 * n_3 = 1$
2. Form first 2 terms $2E_2 = (1493x_1 - 421x_4)^2 * (383x_1 - 108x_4)^2$
3. Add two quadratic terms with $(35x_1 - x_3 - 9x_4)^2$ and $(-10x_1 - x_2 + 4x_3)^2$
4. Replace $x_3 = 35x_1 - 9x_4$ and $x_2 = -10x_1 + 4x_3$ in $2E_2$.

Already in small matrix problems the Gauss elimination process with pivoting can sacrify more than 7 decimals, and a single precision arithmetic leads to a wrong result. Then, 64 bit arithmetic must be used.

Sometimes, single precision arithmetic is used to find an approximate solution that is further improved by an iterative process with double precision arithmetic. A good example is the Linpack implementation in PS3 nodes [29] where single precision is used in the direct solver, followed by the iterative method proposed in [139] applied to double precision operands. This procedure has been proposed to solve linear systems on the first Cell processor that had no hardware double precision arithmetic. Such a mix of single precision direct solver and of a double precision iterative procedure can be less costly and more precise than a direct double precision solver. The complexity of a direct double precision solver for a full matrix of rank N is $\frac{2}{3}N^3$. If single precision arithmetic takes half the time of double precision (this is for instance true for the SSE library: DGEMM takes double time than SGEMM), the complexity reduces to $\frac{1}{3}N^3$, to which one has to add the complexity of the iterative process $N_{it}N^2$, where N_{it} is the number of iteration steps. The CPU times have to be normalized with the performances R_{DGEMM} and R_{DGEMV} for the direct solver and the iterative process, respectively. If $N_{it} < \frac{3R_{DGEMV}}{R_{DGEMM}}N$, the combination of a direct single precision solver with an iterative correction is cheaper than the direct double precision solver. There is a proposal to replace the double precision Lapack matrix solver by such a combination of direct and iterative methods, leading to sometimes faster and more precise results.

5.1.3 Floating point operations

Three floating point operands, x, y, and z are presented in Fig. 5.3. The first bit is the sign of the mantissa, the next bit is for the sign of the exponent, followed by its absolute value, both together form p, q, and r in Eqs. 5.3, and the remaining bits represent the mantissa. The three floating point numbers are

$$x = \pm e \, 2^p$$
$$y = \pm f \, 2^q$$
$$z = \pm g \, 2^r . \tag{5.3}$$

The exponents p, q, and r are positive or negative. The floating point numbers are normalized, i.e. the first bit of the mantissa is always set to 1, and $0.5 \le e < 1$, $0.5 \le f < 1$, and $0.5 \le g < 1$. Let us consider the operation $x + y = z$ (see Fig. 5.4) with $x=1.5$ ($p=1$, $e=0.75$), $y=0.75$ ($q=0$, $f=0.75$) and the result $z=2.25$ ($r=2$, $g=0.5625$).

It is supposed that the add floating point operation is subdivided into 4 suboperations:

		sign	exp	mantissa
Float	x	±	p	e
Float	y	±	q	f
Float	z	±	r	g

Fig. 5.3 The floating point numbers x, y, and z

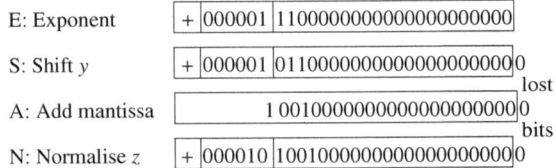

Fig. 5.4 Example of an add operation. One last bit is lost in the shift operation and one bit is lost in the normalization

E. Compare the exponent, i.e. perform $q - p$
S. Shift one of the mantissas by $abs(q - p)$
A. Add the mantissas
N. Normalise the result

Each one of these suboperations is supposed to take one cycle period. In this imaginary example, the add operation takes 4 cycles.

The first step E of the add operation is to adjust the exponent of y to the exponent of x. For this purpose, one computes the shift (p-q=1). In the second step S, the mantissa of y is shifted by 1 to the right, $f/2=0.375$, 1 is added to q, and $q + 1=p$. In the third step A, the two mantissa are added, giving $g=0.75+0.375=1.125$. The normalization step N demands a shift of the mantissa g by 1 to the right, and 1 is added to the exponent r. After normalization, the mantissa is 0.5625, and the result $4 * 0.5625 = 2.25$, as expected.

We have to note here that one loses the last bits when shifting a mantissa to the right. As long as the lost bits are zeros, there is no loss in precision. The underflow bit is handled in a statistical manner, sometimes it is doubled, sometimes just lost. This reduces the loss in precision. Even though, the truncation errors can lead in loosing all the decimals, as shown before with the matrix problem, Fig. 5.2.

5.1.4 Pipelining

In old von Neumann machines that did not include pipelining, an arithmetic operation took many cycle times. Only one step of the set of sub operations was in execution at a time, and each step took many cycle periods.

All the processors now on the market have adopted pipelining of data transport from and to main memory, of data transport between the caches and the register, of instruction interpretation, of computation, and of I/O operations. Let us demonstrate pipelining of the add operation. There are four suboperations that are independently hardware implemented and arranged in a pipeline.

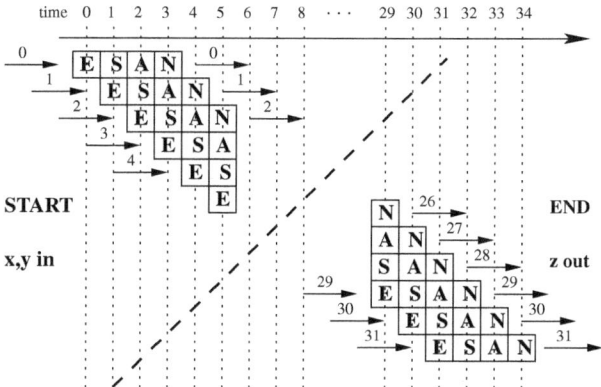

Fig. 5.5 The pipelined add operation

For the vector add operation
for $i = 1, 32$ **do**
   $z(i) = x(i) + y(i)$
end for
it is possible to activate all these suboperations simultaneously. This is shown in
Fig. 5.5. In the upper part of the Figure are indicated the cycle times from 0 to 34.
At the first cycle period the first operation $z(1) = x(1)+y(1)$ enters into the pipeline
with the suboperation E. During the second cycle period, the second step S of the
first operation is executed, whereas the second operation $z(2) = x(2) + y(2)$ starts
with suboperation E. After 4 cycle periods, the first result, $z(1)$, leaves the pipeline,
the second operation enters step N, the third operation enters step A, the fourth
operation enters S, and the fifth operation just enters the pipe, step E. From now
on, one result per cycle period comes out the pipeline. The whole loop with 32 adds
takes 35 cycle periods, 3 cycles for filling the pipeline, and 32 cycles during which
one result is produced per cycle period.

   The pipelining does not stop with the add operation. It is possible to enter the
result from an add pipeline directly into a multiply pipeline, or, as it appears more
often, a result coming from a multiply pipeline directly enters an add pipeline.
This happens most efficiently if the other operands of such pipelined operations are
located in cache, as in BLAS3 and FFT operations. We have to mention here that an
add and a substract operation take the same time. This is not the case for multiply
and divide operations. Multiply has a higher latency than an add instruction, and a
divide takes much longer and should, if possible, be replaced by multiplying with
its inverse.

   Pipelining is comparable with a production line in which each production step
takes exactly the same time, i.e. one cycle period. Such a pipeline is globally slowed
down if one step is slowed down. For instance, if the vector add operation is too

long, and data has to be brought from main memory, the pipeline has to wait for data, and is slowed down.

5.2 Single core optimization

Single core code optimization has to do with main memory conflicts and with increasing cache hits when running main memory access dominated applications. Most modern compilers are able to minimize the number of cache misses and to recognize main memory conflicts and correct them, "hand optimization" examples are presented to understand how performance can be improved.

5.2.1 Single core architectures

Let us explain in more detail how data moves in a single processor, single core node as in the architecture of an Intel Pentium 4 node (Figure 5.6) with a cycle period of 2.8 GHz and a peak of R_∞=5.6GF/s. There are two functional units, an add and a multiply pipe. The node is connected to a motherboard with a Northbridge chip (that connects to the two DDR memory units and to the graphical card) and Southbridge chip (that connects to I/O and to the communication networks). In the case of a cache miss, an operand needed by the processor has to be transferred from main memory to the highest level cache, i.e. for a Pentium 4 to level 2 cache, with a peak bandwidth of 6.4 GB/s. The bandwidth from level 2 cache through level 1 cache to the processor registers is 44.8 GB/s, i.e. two 64bit word or four 32bit words per cycle period. The difference between the two caches is latency time: level 2 cache has a latency of 7 cycle periods, whereas the level 1 cache has a latency of between 2 to 6 cycle periods. There are 2 GB of slow DDR main memory, the cache level 2 with its relatively long latency has a size of 1 MB, where the low latency level 1 cache has only 8 kB of local memory. This demonstrates that data in cache transits data at least 7 times faster to the processor than from main memory to cache. The smaller g_m (we remind that g_m=0.5 for the Pentium4) is, the bigger this factor, here, V_m=7/g_m=14. In addition, cache memory access does not suffer much from non-stride 1 data positioning, but main memory access does, as we will see in the next section.

5.2.2 Memory conflicts

To explain main memory access conflicts, let us come back to the matrix times matrix multiplication $C = A \times B$. The matrices A, B, and C are all full. The typical triple loop is shown in Fig. 5.7 at the left where the matrix element $c(j, k)$ is computed by summing up the scalar product of row i of $a(j, i)$ times the i column of $b(i, k)$. This triple loop is executed on an Intel Pentium 4 single core node running at 2.8 GHz. The lowest possible compilation option -O1 has been chosen to make

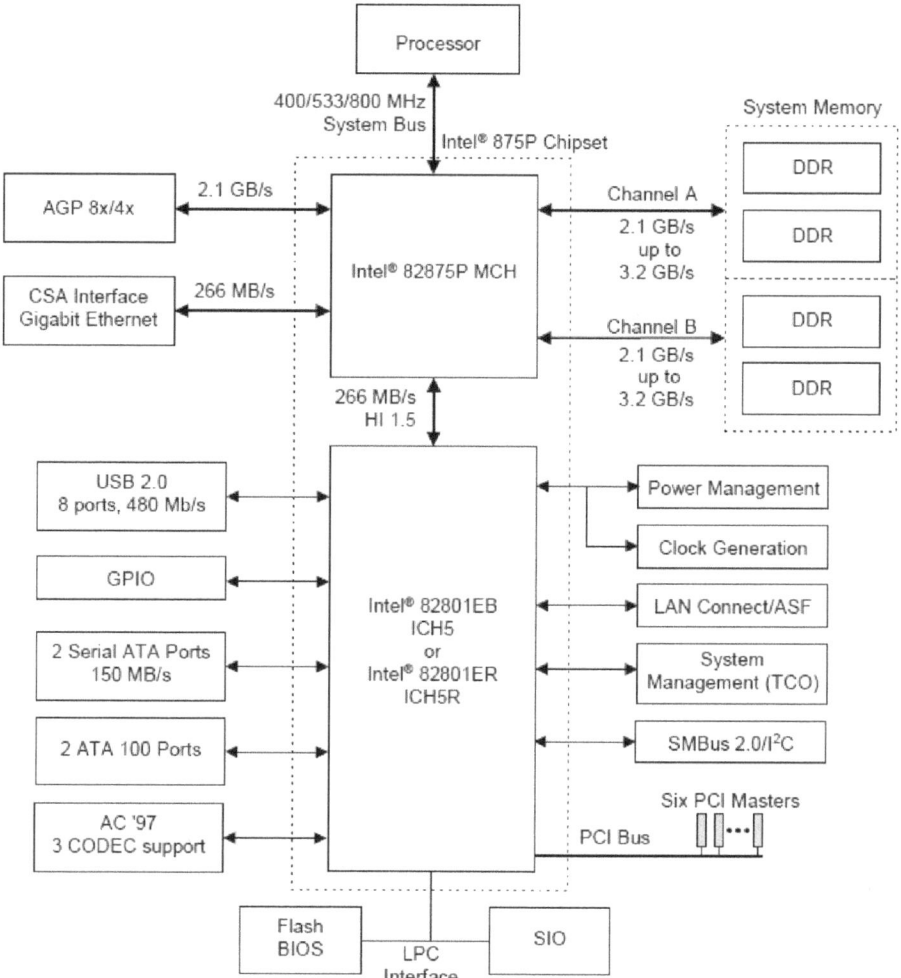

Fig. 5.6 The architecture of a Pentium 4 computer in 2005. There is a North chip connected to the processor, to the main local memory (DDR), to the graphics card, and to the South chip that also connects to the I/O units

main memory conflicts visible. The measured performance $R_a = 0.196$ GF/s is very small with respect to the 4.8 GF/s obtained when replacing this triple loop by a call to the BLAS3 subroutine DGEMM written in assembly language to get the highest possible processor performance.

In the initial triple loop the rows of $a(j, i)$ have to be brought to cache for all loop indices i and j. The $c(j, k)$ element is constant in the innermost loop, i.e. for all i. For $n = 1000$ as indicated in Fig. 5.7, the vector $b(i, k)$ has a length of 1000 words (=8000 Bytes), and can be kept in cache for all j. For each $a(j, i)$ element brought to cache, there are two operations done (a multiply and an add), i.e.

```
for k = 1 to n do                          for k = 1 to n do
   for j = 1 to n do                           for i = 1 to n do
      for i = 1 to n do                            for j = 1 to n do
         c(j,k) = c(j,k)+a(j,i)*b(i,k)                c(j,k) = c(j,k)+a(j,i)*b(i,k)
      end for                                      end for
   end for                                      end for
end for                                      end for
```
$n = 1000, V_a = 2, R_a = 0.196\,\text{GF/s}$ $n = 1000, V_a = 2, R_a = 0.772\,\text{GF/s}$

Fig. 5.7 Left: Standard triple loop for a matrix times matrix multiplication leads to main memory conflicts. **Right**: After exchange of loops in i and j, conflict has disappeared. Performance is now three times bigger than expected

$V_a = 2$. With $M_\infty = 0.8Gw/s$ and $g_m = 0.5$, one would predict a performance of $R_a = g_m * V_a * M_\infty = 0.8GF/s$, but only 0.196 GF/s are measured.

The problem originates from what is called the main memory conflict. If data in memory is not accessed contiguously, i.e. with a stride=1, the memory bandwidth drops. This is due to the "paging" mechanism to read chunks of 64bytes that are stored consecutively in main memory. If one reads a matrix row by Fortran or a column by C, the operand addresses jump in main memory. In fact, in the left hand side triple loop in Fig. 5.7 two consecutive elements in the row of matrix $a(j,i)$ have a distance in memory of 8*n Bytes. As a consequence, main memory bandwidth drops by a factor of 4 due to constantly occurring cache misses.

This can quite easily be corrected by interchanging the i and j indices. The new triple loop shown at the right in Fig. 5.7 gives the same result as the one at the left. The operations are done in another sequence. Now, $b(i,k)$ is constant for all j, and $c(j,k)$ remains in cache for all i. The innermost loop index j is now over the columns of matrix A, the access is done with stride=1, and the measured performance $R_a = 0.772$ GF/s is close to the expected $R_a = 0.8$ GF/s. We have to mention here that the higher optimization compiler options (O3 for instance) automatically perform an exchange between the i and the j loops. The measurements presented in Fig. 5.7 have been made with an ifc 7.0 compiler.

This example was just to learn how main memory conflicts can lead to poor performance when the compiler does not correct them. In fact, if we measured the performance of the loops on a Pentium 4 running at 2.8 GHz with the Fortran compiler options -O3 -xW, both loops give the same performances, the compiler automatically exchanged the loops. The measures values, however, are much larger than what we would have expected. DGEMM delivers 5 GF/s (out of 5.6 GF/s peak), independent of the compiler version. This 89% of peak performance is due to the assembler written coding by Goto. Using ifort 9.1 and without DGEMM, the two loops deliver 0.97 GF/s, slightly higher to the 0.8 GF/s one would expect. The surprise comes with ifort 10.0. With this compiler, the performance increases to 2.6 GF/s. This corresponds to more than half of DGEMM. The ifort 11.0 compiler slightly reduces the performance to 2.3 GF/s.

On a single processor, 4 cores Nehalem running at 2.7 GHz, DGEMM uses, i.e. on peak performance. Th reach this unbelievable performance, it uses the SSE library that can deliver up to 4 results per cycle period, with a peak of 10.8 GF/s. In Table 6.1 we see that DGEMM runs at 10.7 GF/s. This ultra high performance is due to the overclocking of the nodes, called Turbobooster. With the options -O3 -xSSE4.2 the ifort 11.1 compiler delivers 7.8 GF/s for both loops. This is over 70 % of SSE peak, and larger than core peak.

Another example to reduce cache miss is the sparse matrix times vector operation SMXV (see Fig. 3.3). The multiplication of the three tri-diagonal blocks with the vector can be done in three different ways:

Figure 5.8: the SMXV is realized by 9 vector times vector operations. The $V_a = O/M = 0.5$ value is given by two operations ($O = 2$) and 4 main memory accesses ($M = 4$). This leads to a performance of $R_a = 0.2$ GF/s.

Figure 5.9: the SMXV is realized by 3 tri-diagonal matrix times vector operations. The $V_a = O/M = 1$ value is given by six operations ($O = 6$) and 6 main memory accesses ($M = 6$). In fact, there are loads for the tri-diagonal matrix, one load for the vectors **u** and **ua** each, and one store for **ua**. This leads to a performance of $R_a = 0.4$ GF/s. This test case has an exact value of $V_a = 1$ and is used to determine experimentally the g_m values of different cores.

Figure 5.10: the SMXV is realized by one matrix times vector operations. The $V_a = O/M = 17/12$ value is given by 17 operations ($O = 17$) and 12 main memory accesses ($M = 12$). In fact, there are loads for the nine matrix elements, one

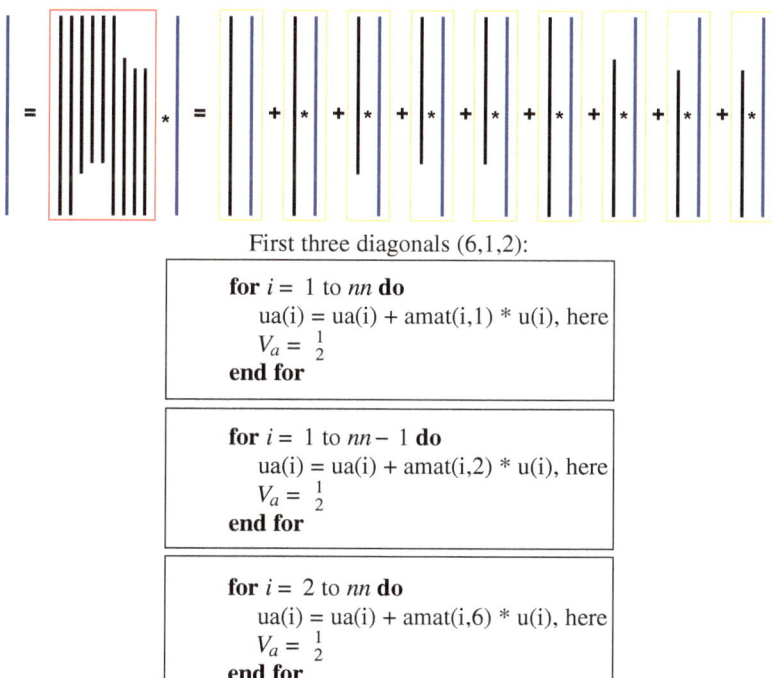

First three diagonals (6,1,2):

for $i =$ 1 to nn **do**
    ua(i) = ua(i) + amat(i,1) * u(i), here
    $V_a = \frac{1}{2}$
end for

for $i =$ 1 to $nn-$ 1 **do**
    ua(i) = ua(i) + amat(i,2) * u(i), here
    $V_a = \frac{1}{2}$
end for

for $i =$ 2 to nn **do**
    ua(i) = ua(i) + amat(i,6) * u(i), here
    $V_a = \frac{1}{2}$
end for

Fig. 5.8 Vector by vector multiply and add. Nine vector times vector multiplies

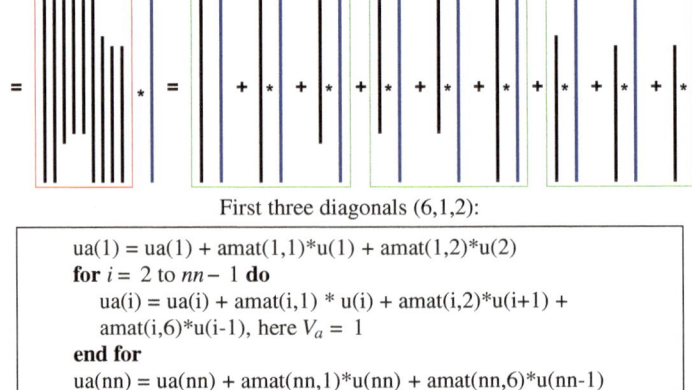

First three diagonals (6,1,2):

```
ua(1) = ua(1) + amat(1,1)*u(1) + amat(1,2)*u(2)
for i = 2 to nn− 1 do
    ua(i) = ua(i) + amat(i,1) * u(i) + amat(i,2)*u(i+1) +
    amat(i,6)*u(i-1), here Va = 1
end for
ua(nn) = ua(nn) + amat(nn,1)*u(nn) + amat(nn,6)*u(nn-1)
```

Fig. 5.9 Triangular matrices times vector multiplies and add

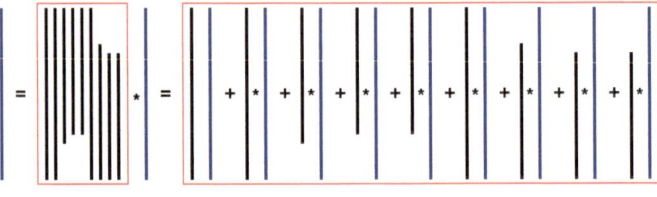

```
Lines n1+3 to nn-n1-2 :

jc = n1 + 1
for i = jc+ 2 to nn− jc− 1 do
    ua(i) =
    amat(i,9)*u(i-jc-1) + amat(i,8)*u(i-jc) + amat(i,7)*u(i-jc+1) +
    amat(i,6)*u(i-1) + amat(i,5)*u(i) + amat(i,4)*u(i+1) +
    amat(i,3)*u(i+jc-1) + amat(i,2)*u(i+jc) + amat(i,1)*u(i+jc+1) +
end for
ua(nn) = ua(nn) + amat(nn,1)*u(nn) + amat(nn,6)*u(nn-1)
```

Fig. 5.10 One matrix times vector multiply

load for the vectors **u** and **ua** each, and one store for **ua**. Note that the accumulation of the result into ua has not to be done. This in future will be the best case. Today's compilers give a performance comparable with the $V_a = 1$ case.

5.2.3 Indirect addressing

The same increase in main memory access time can be observed if we have an indirect addressing Fortran instruction such as

$$c(ind(j), k) = c(ind(j), k) + a(ind(j), i) * b(i, k). \tag{5.4}$$

The indirect addressing with $ind(j)$ leads to jumps in main memory and to cache misses if the sparse matrices C and A cannot be kept in cache. It would be advantageous if the indirect addressing would be interchanged with the second k counter.

5.2.4 Unrolling

In this subsection is presented the fundamental idea of unrolling an outer loop. Since the compilers do unrolling automatically, it is not advisable to do unrolling in the code. It drastically reduces readability and maintenance capabilities. In addition, when coding up unrolling, the compiler could get a problem to best optimize. By presenting loop unrolling, we can learn how the performance can be increased.

We have learned that the number of operations per main memory access, V_a, has to be as high as possible. In Fig. 5.11 it is shown how V_a can be increased by a factor of 2 just by unrolling the triple loop. On its left is shown the same loop as in Fig. 5.7. Jumping the most outside k-loop by two demands a correction of the innermost j loop. There are now two instructions, one for the index k, and one for the index $k + 1$. What did we gain?

Now there are two constants, $b(i, k)$ and $b(i, k + 1)$, and two vectors, $c(j, k)$ and $c(j, k + 1)$ that are kept in cache. The column of $a(j, i)$ now appears in both instructions. For each load operation of $a(j, i)$ it is now possible to perform 4 operations (2 multiplies, 2 adds), and $V_a = 4$. The consequence is that the performance $R_a = 1.342$ GF/s has almost doubled. If the number of unrolled operations is doubled again, another performance increase is obtained. Theoretically, this should be possible until the cache is full. We have to mention here that the unrolled loop has to be completed if the matrix rank n is not dividable by the unrolling level 2.

Here we have to mention that the compiler optimization options include an optimal unrolling of the relevant operations. It is even able to access the SSE hardware. Therefore, unrolling in the code can be contra-productive, and should never be done.

```
for k = 1 to n do
    for i = 1 to n do
        for j = 1 to n do
            c(j,k) = c(j,k)+a(j,i)*b(i,k)
        end for
    end for
end for
n = 1000, Va = 2, Ra = 0.772 GF/s
```

```
for k = 1 to n, 2 do
    for i = 1 to n do
        for j = 1 to n do
            c(j,k) = c(j,k)+a(j,i)*b(i,k)

            c(j,k+1) = c(j,k+1)+a(j,i)*
            b(i,k+1)
        end for
    end for
end for
n = 1000, Va = 2, Ra = 1.342 GF/s
```

Fig. 5.11 Almost a factor of 2 increase of R_a is measured when unrolling is performed. **Left**: best loop after exchanging the indexes due to memory conflicts (see Fig. 5.7). **Right**: unrolled kij loop (that must be corrected for odd n)

5.2.5 Dependency

Dependency inhibits pipelining of the operations. This leads to a performance reduction by one order of magnitude, dependencies should therefore be eliminated. Such a case is illustrated by the Gauss-Seidel iteration used to solve iteratively a system of linear equations from finite difference and finite element approximation methods. In Fig. 5.12 the original Gauss-Seidel iteration for a two-dimensional structured finite difference method is shown. The new green central circle solution is computed by using the surrounding red square values. The computation goes from left to right and from bottom to top. The new green value can only be computed after having evaluated the left side red one, and pipelining is stopped. As a consequence, the performance is very low.

Figure 5.13 shows an iterative algorithm in which the mesh points are differently numbered, or colored. First, one numbers the red points, then the black points. All red points are surrounded by black points, and all black points are surrounded by red points. One first computes for the red points, fixing the values of the black ones. Then, one fixes the values of the red points, and computes the black ones. The original Gauss-Seidel iteration and the red-black scheme converge similarly well. The red-black iteration is much faster than the original one.

In a two-dimensional finite element approach on a structured mesh, four colors have to be chosen to get rid of dependency, see Fig. 5.14.

$$\nabla^2 u = \frac{\partial^2 u}{\partial x^2} + \frac{\partial^2 u}{\partial y^2} = 0, \in \Omega$$

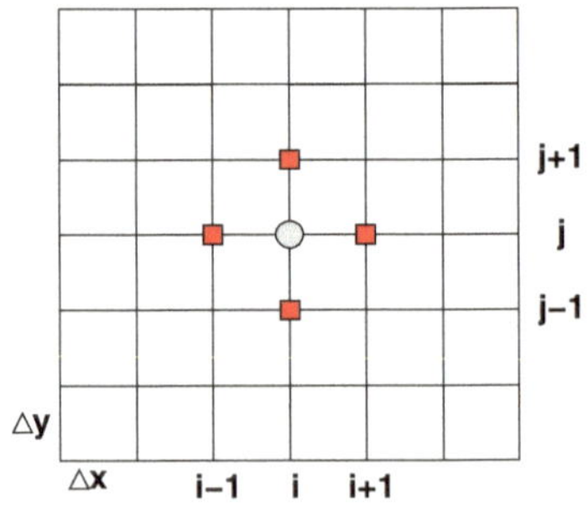

Approximation : $\dfrac{u_{i-1,j} - 2u_{i,j} + u_{i+1,j}}{\Delta x^2} + \dfrac{u_{i,j-1} - 2u_{i,j} + u_{i,j+1}}{\Delta y^2} = 0$

Scheme : $\dfrac{(u(i-1,j) - 2u(i,j) + u(i+1,j))}{\Delta x^2} + \dfrac{(u(i,j-1) - 2u(i,j) + u(i,j+1))}{\Delta y^2} = 0$

Solve for $u(i,j)$ **and advance to the right**

Fig. 5.12 Dependency: Mesh point (i+1,j) can only be computed after mesh point (i,j)

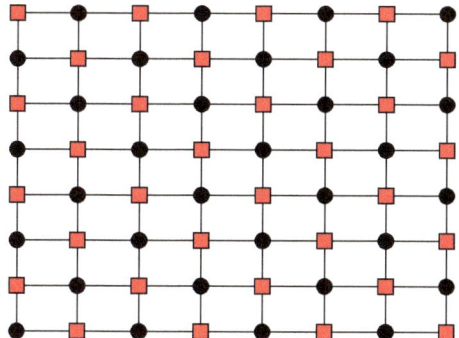

Compute the red points :

```
for k = 1 to nit do
    for j = 1 to m do
        istart = j − 2 (j − 2)/2
        for i = istart to n, 2 do
            u(i, j) =
            (u(i − 1, j) + u(i + 1, j))/qx+
            (u(i, j − 1) + u(i, j + 1))/qy
        end for
    end for
```

Compute the black points :

```
for j = 1 to m do
    istart = j − 2 (j − 1)/2 + 1
    for i = istart to n, 2 do
        u(i, j) =
        (u(i − 1, j) + u(i + 1, j))/qx+
        (u(i, j − 1) + u(i, j + 1))/qy
    end for
end for
```

Fig. 5.13 Red-Black iteration: All red points then all black points can be computed. That method converges as fast as Gauss-Seidel. Get to stride 1 by sorting : Number first all the red points. Then number all the black points. $qx = (dx^2 + dy^2)/dy^2$, $qy = (dx^2 + dy^2)/dx^2$

In an unstructured finite element method, the number of colors can even be higher, Fig. 5.15. But in any case, it is always possible to choose a coloring such that all the nearest neighbors have a different color than the variable to be computed.

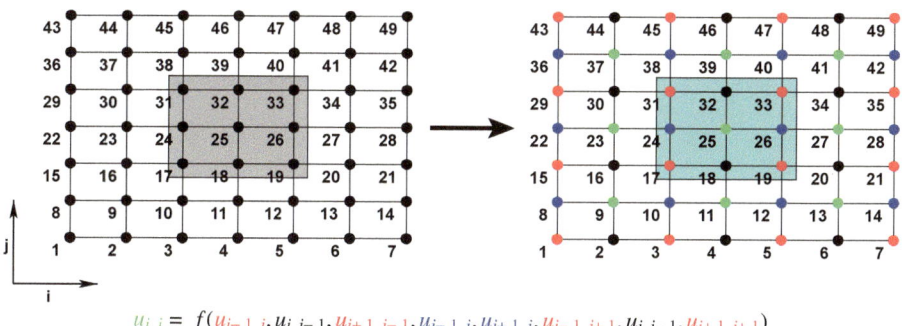

$$u_{i,j} = f(u_{i-1,j}, u_{i,j-1}, u_{i+1,j-1}, u_{i-1,j}, u_{i+1,j}, u_{i-1,j+1}, u_{i,j-1}, u_{i+1,j+1})$$

Fig. 5.14 Four color iteration for a structured finite element approach. Finite element coloring for Laplacian ($m = 7$, $n = 7$, $b = 0$, $\Delta x = \Delta y = 1$) on a structured mesh

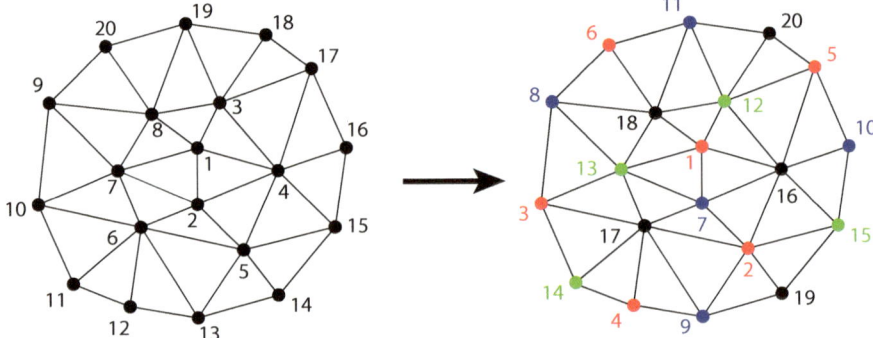

Fig. 5.15 Four color iteration for an unstructured finite element approach

5.2.6 Inlining

In a program the following call to a small subroutine is found in a loop

```
for i = 1, N do
    call SAXPY(...)
end for
```
Algorithm 1: Call of a subroutine within a loop leads to low efficiency

This call to the BLAS1 SAXPY kernel interrupts the instruction pipeline, and stops the pipelined execution in the loop. If the call to SAXPY is replaced by introducing the instructions of the kernel directly into the loop, pipelining is activated. Compiler options exist to do so. This increases efficiency of the code.

5.2.7 If statement in a loop

A very similar problem is an IF statement in a loop.

```
for i = 1, N do
    if i.eq.N then
        foo()
    else
        ber()
    end if
end for
```
Algorithm 2: Problem with an IF statement within a loop

```
for i = 1, N − 1 do
    ber()
end for
foo()
```
Algorithm 3: Same than Alg. 2 but without an IF in the loop

This was simple, and the compiler would have automatically ignored the IF, and corrected afterwards. In more complicated cases, branch prediction is made, and corrected if prediction was wrong.

5.2.8 Code porting aspects

When porting codes, especially those with a long history, one should be prepared for surprises. All too often one finds optimizations that were a good match for earlier generations of computers, but that prove counterproductive on current hardware. There are also software constructs better expressed using more advanced language features, such as Fortran loops replaceable with array expressions. And then one finds source code that was never really part of the language used to write a program.

As an example, we present pseudo code modeled on parts of a subroutine ported through several generations of scalar and vector processors to the Pleiades cluster at EPFL. The quantities ns, nz, $imax$, and $mmax$ are integer parameters giving array dimensions and loop limits. n, i, m, and mp are loop indices. Arrays have dimensions

```
real, dimension(ns,nz) :: q1, q2
real, dimension(ns*nz) :: r1, r2

real, dimension(ns*nz,imax,2) :: xr, xz
real, dimension(imax, mmax) :: ci, cm, si, sm
```

The following loop appears as the innermost loop of a loop nest, with the middle and outer loops using the variables m and j.

> **do** $n = 1, ns * nz$
>   q1(n,1) = q1(n,1) + r1(n)*ci(i,m) + xr(n,i,mp)*sm(i,m)
>   q2(n,1) = q2(n,1) + r2(n)*si(i,m) + xz(n,i,mp)*cm(i,m)
> **end**

Unfortunately, $q1$ and $q2$ are defined as two-dimensional arrays, and used as such in other parts of the subroutine. Here the programmer is attempting to simulate collapsing a two-dimensional loop nest into a single loop by letting the loop index vary, not only through the first column of arrays $q1$ and $q2$, but throughout all of arrays $q1$ and $q2$. So the index n exceeds the limit ns allowed for the first subscript of $q1$ and $q2$ if nz is greater than one (which it happens to be).

The only problem is that this is prohibited by the Fortran language specification (see Section 6.2.2.1 of Fort 95 [83]):

The value of a subscript in an array element shall be within the bounds for that dimension.

Section 5.4.2 of the earlier Fortran standard [Fort77] contains the same requirement, although the wording is different.

To be certain of the meaning of Section 6.2.2.1, please see Section 1.6 of [83], Notation used in this standard:

> In this standard, "shall" is to be interpreted as a requirement; conversely, "shall not" is to be interpreted as a prohibition. Except where stated otherwise, such requirements and prohibitions apply to programs rather than processors.

The reason for requirements such as Section 6.2.2.1 of [83] is that by not attempting to give any interpretation whatsoever to broken code, the compiler is often free to make correct code run much faster than would otherwise be possible. It is the programmer's responsibility to make certain that these requirements are met.

Exactly what happens when Section 6.2.2.1 is not met is deliberately left undefined by the standard. Sometimes the generated object code does exactly what the programmer intended; sometimes it doesn't. The program is in that delightful state of quantum uncertainty so well described by Wolfgang Pauli so many years ago:

> "This isn't right. It isn't even wrong."

As a very much simplified example, imagine the following code sequence:

```
real x,y(10,2)
integer i,n
...
DEFINE y ! Give some values to elements of array y
DEFINE n ! Give some value to integer n
...
do i = 1,n
 y(i,1) = y(i,1) * 3.0
end do
x = sqrt(y(10,2))
```

Now imagine that the programmer defines n as 20 in the expectation that the generated object code will sweep through both columns of y, each of length 10, with column 2 stored right after column 1 in memory. The compiler realizes, however, given the logical implications of Section 6.2.2.1 of [83], that the DO-loop is allowed to access and modify only the first column of the array y. The second column, and therefore the value of $y(10, 2)$, cannot be changed by this DO-loop according to the rules of the language. The two operations (the DO-loop and the square root) are completely independent.

Therefore, the compiler can increase performance by performing the two operations in parallel. The generated object code could:

1. Load $y(10, 2)$ into a machine register
2. Initiate a $SQRT$ instruction
3. Begin the DO-loop multiplying the first column of array y (or as many elements of the first column of array y are indicated by the value of n) by 3
4. Store value of completed $SQRT$ instruction in scalar variable x

In the best of all possible worlds, the $SQRT$ instruction finishes execution at exactly the same time as the sweep through the DO-loop. Both computations overlap, giving a factor of two speedup on this code segment.

The only problem is that this sequence isn't exactly what the programmer wanted. The compiler, however, is working perfectly according to the definition of the Fortran language.

Is anything like this going on inside the program in question as ported to Pleiades? Unless a complete static analysis of the generated object code is done, or unless every input data set is run through a correct version of the program and the output compared with that of the misoptimized version, nobody knows.

Can we eliminate such problems with broken source code by compiling only with reduced optimization? In a word: No. Even the most basic code generation algorithms can produce object code that does something quite different from the programmer's intent if the source code violates the rules of the language. Imagine for a moment the following subroutine:

```
subroutine scale3(x,n1,n2)
implicit none
real, intent(inout), dimension(:,:) :: x
integer, intent(in)                  :: n1,n2
integer                              :: i
do i = 1,n1*n2
 x(i,1) = x(i,1) * 3.0
end do
end subroutine scale3
```

Further imagine that subroutine *scale3* is contained in another routine, a bit of which appears here:

```
. . .
real, dimension(4,2) :: y
. . .
. . . ! Code assigning values to all elements of array y
. . .
call scale3 (y(1:3,1:2), 3, 2)
. . .
```

That is, the containing routine calls *scale3*, not with an entire array as the dummy argument *x*, but with an array section with part, but not all, of the 8-element two-dimensional array *y*.

If subroutine *scale3* were an external routine, the compiler-generated object code would copy the array section $y(1 : 3, 1 : 2)$ into a contiguous area of memory, call *scale3* with this area as the array argument, then copy it back after the call. Since, however, *scale3* is contained in the calling routine, the calling routine has an explicit interface to *scale3*. Information passed between caller and callee is no longer limited to call-by-reference or copy-in copy-out. More complete information exchange is now both allowed and potentially useful.

Hence the compiler might choose to pass a dope vector describing the array section argument, specifying the starting address of y, its extent in each dimension, and the extents of the array section used as the actual argument. Internal to $scale3$, the generated object code would calculate the actual address of element $x(i, j)$ of the dummy argument array x as

```
$base_address$ + (i-1) + 4*(j-1)
```

with the i and j-dependent modifiers scaled by an additional factor of four, assuming that real variables are thirty-two bits in length and the computer in question addresses to the octet level.

As a deliberately bad example, we've miswritten the DO-loop inside subroutine $scale3$ in gross violation of Section 6.2.2.1 of the Fortran language specification. Here i runs from 1 to 6; j remains 1. The generated object code using the dope vector will modify elements

```
y(1,1), y(2,1), y(3,1), y(4,1) [Wrong!],
        y(1,2), and y(2,2)
```

and fail to modify $y(3, 2)$ [Also wrong].

The compiler might choose to use dope vectors even with the lowest level of optimization. Indeed, it may choose to do so especially with the lowest level of optimization.

As another example, consider a large CFD program at a previous place of employment. After two years of running on Machine A, the users attempted to port their program to the newly-installed Machine B. Machine B included a new and more aggressive Fortran compiler. It also gave different results than Machine A.

As eventually established, the calculations began to diverge in a subroutine containing code something like this:

```
real, dimension(10,20,30) :: x, y, z
integer i,j,k
...
do k = 1, 30
  do j = 1, 10
    do i = 1, 20
      x(i,j,k) = y(i,j,k) + alpha*z(i,j,k)
    end do
  end do
end do
```

Note the limits on the inner and middle DO-loops. Somewhere, sometime, some-how, someone switched the upper limits on the *i* and *j* loops. Two years of wrong answers.

On Machine *A*, the generated object code happened to do what might be con-sidered the most obvious and swept through two columns rather than one on each sweep through the innermost loop. Machine *B*, being more aggressive, unrolled the middle loop, to a depth of 4, if memory serves. As a result, memory references that shouldn't have overlapped, but did, occurred in a different order than on Machine *A*, giving a different set of wrong answers.

Could a compiler switch to generate runtime code to check for out-of-range sub-scripts have found this problem? Alas, not as the program was written. Besides the accidental violation of Section 6.2.2.1, the whole source code was full of DO-loops such as

```
do ijk = 10*20*30
  x(ijk,1,1) = 0.0
end do
```

Just think of installing smoke detectors in a building where everybody smokes cheap cigars.

Moral: The only way to deal with broken source code is not to have any.

5.2.9 How to develop application software

In fact, this section has been compiled from real-life experiences of a prominent software engineer who helped supercomputer users to port their codes. it has been added to show what can happen if one has to port legacy or broken codes from one computer architecture to another one. Not to have any broken codes is to write programs that follow strictly clearly defined rules. The codes should be modu-larly organized. Some computer engineers are happy with the object oriented C++ programming language. A good example is the SpecuLOOS code presented in chapter 8. Others use data bases to ease data transfer from one application com-ponent to another one. Such example can also be found in chapter 8.

Code development should be performed by teams including scientists to define the physics models, numerical mathematicians to define the solvers, and computer scientists to implement the numerical methods most efficiently on cores, nodes, and clusters. Results have then to be interpreted by the scientists who are specialists in the research domain. Code validation and data interpretation are arts. Specifically, it is not easy to find errors in a code. If there is an error, the research scientist has the tendency to interpret wrong results as to be physically correct. We definitely need independent code tester who always believes that the results are wrong, as long as he is not convinced from the contrary. He should try to find another code to compare with, or rewrite a same code sequence with another method. In those cases, it is needed to know how to interpret the results of both codes. If the results do not coincide, one or both codes are wrong, if they coincide, the codes give a probably right result, but only for this specific input case.

5.3 Application to plasma physics codes

The optimization procedures are demonstrated with the two applications in thermonuclear fusion research, VMEC and TERPSICHORE, both ran perfectly well on the vector NEC SX Series. Other examples can be found in [57]. We shall see that porting an application from a vector machine to a PC, or vice-versa, demands an adaptation of the program to the new computer architecture.

5.3.1 Tokamaks and Stellerators

The thermonuclear fusion research deals with a hot ionized gas, called plasma, that is kept away from material walls by means of strong magnetic fields. The most popular fusion machine is the so-called Tokamak, the ITER fusion reactor experiment Fig. 5.16 is such an installation. In a Tokamak the plasma is confined by a

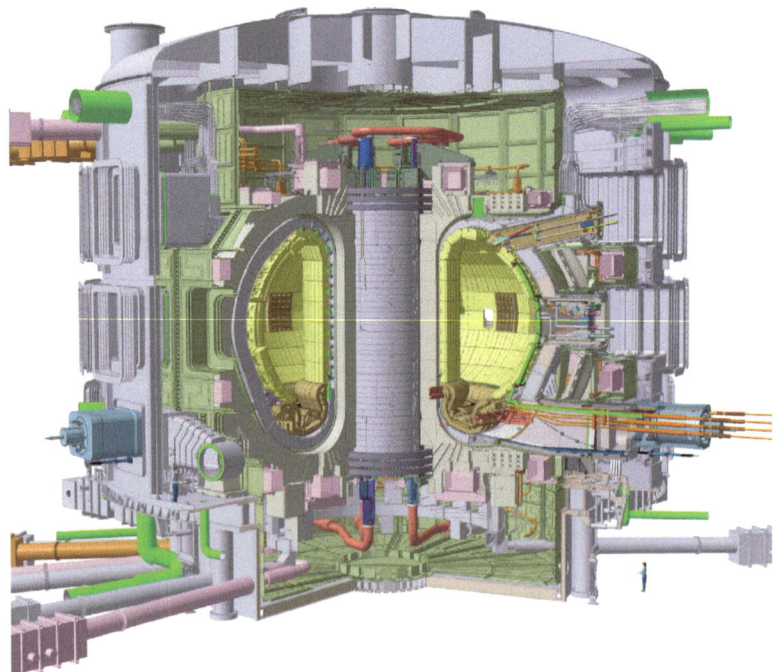

Fig. 5.16 ITER: The International Toroidal Experimental Reactor

huge toroidal magnetic field produced by magnetic field coils, superimposed by a
poloidal magnetic field induced by a toroidal plasma current. The equilibrium position of the magnetic flux function (see Fig. 5.19) is described by the non-linear ideal
two-dimensional (2D) magneto-hydrodynamic (MHD) equilibrium equations [66].
To find out if the equilibrium is stable or unstable, the equilibrium solution is perturbed, and the MHD equations linearized. If the resulting eigenvalue problem has
no negative eigenvalues, the configuration is stable, a negative eigenvalue reveals
existence of an instability that destroys the equilibrium in microseconds time scale.
With the stability code ERATO [68], named after the Greek muse of music, song and
dance, the "scaling law" of Tokamaks has been found [131], and later confirmed by
experimental measurements. The KINX [42] code can also treat axisymmetric plasmas with an X-point needed in burning plasmas to eliminate the produced Helium
ions.

In a Stellarator, the torus is deformed to a three-dimensional (3D) helical geometry as the Wendelstein 7X stellarator shown in Fig. 5.17. The equilibrium configuration is periodic along the toroidal angle. The equilibrium is found solving the
nonlinear equations

$$\nabla p = \mathbf{j} \times \mathbf{B} \tag{5.5}$$
$$\mathbf{j} = \nabla \times \mathbf{B}$$
$$\nabla \cdot \mathbf{B} = 0,$$

with $p, \mathbf{j}$, and $\mathbf{B}$ being the pressure, current density, and magnetic field, respectively. These equations are restricted by boundary conditions that fix the position
of the plasma column, Fig. 5.17. Computing numerically this 3D non-linear MHD
equilibrium position is difficult, and from a mathematical point of view, the existence of embedded magnetic surfaces without magnetical islands, needed for ideal
MHD stability studies, has not been demonstrated. Nevertheless, such equilibria are
forced, and a program, called VMEC, has been written for this purpose [76]. The
variable poloidal magnetic flux function is expanded in Fourier series in the two
angular directions, and finite differences are used in "radial" direction. VMEC can
also be used to compute ideal 2D equilibria.

Such ideal MHD equilibrium solutions are perturbed, and the linearized MHD
equations are treated by TERPSICHORE [5], named after the Greek muse of dance
and lyric poetry. These equations can be written in variational form,

$$\delta W_p + \delta W_v - \omega^2 \delta W_k = 0, \tag{5.6}$$

where δW_p is the potential plasma energy, δW_v is the magnetic vacuum energy
around the plasma, δW_k is the kinetic energy, and ω^2 the eigenvalue of the system. A magnetically confined configuration is ideal MHD stable if all eigenvalues
are positive. One single negative eigenvalue reveals an instability. The sign of the
smallest eigenvalue tells us if the configuration is stable or not, and, if negative, the
configuration is unstable.

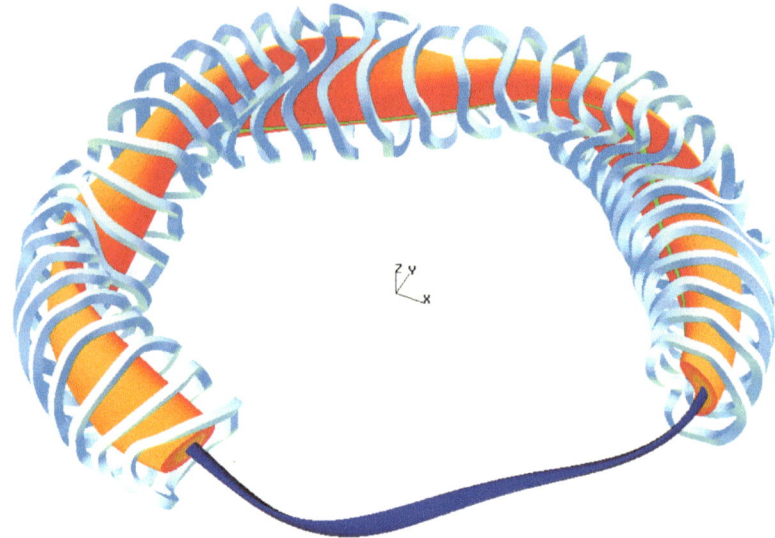

Fig. 5.17 Model representation of the Wendelstein 7X stellerator installed at Max Planck Institut für Plasmaphysik in Greifswald, Germany. The deformed magnetic field coils produce the helical magnetic field wit 5 periods. There is no net toroidally induced current for the poloidal magnetic field. In red and blue are shown two magnetic flux surfaces. The green line on the red surface is a magnetic field line around which charged particles rotate

```
 %cum       self   self tot      times
38.25     707.82    707.82       3130   0.23   0.23   totzsp_.old
32.14    1302.69    594.87       3124   0.19   0.19   totmnsp_.old
 8.78    1465.20    162.51       3124   0.05   0.05   forces_
 7.05    1595.58    130.38       3125   0.04   0.04   bcovar_
 3.95    1668.61     73.03       3125   0.02   0.02   getiota_
 2.39    1712.82     44.21       3128   0.01   0.01   alias_
 2.32    1755.82     43.00       3130   0.01   0.01   jacobian_
 1.96    1792.14     36.32       3130   0.01   0.01   funct3d_
 0.82    1807.39     15.25     339277   0.00   0.00   sdot_
 0.79    1822.08     14.69        284   0.05   0.05   precondn_
 0.56    1832.51     10.43      12496   0.00   0.00   trid_
```

Fig. 5.18 Initial profiling of VMEC

In the ideal linear MHD equations charged particles move along magnetic field lines, and rotate around them. They never leave a magnetic flux surface. To satisfy this fundamental condition, a special coordinate system is constructed in which the magnetic flux surfaces are the new "radial" coordinates as shown in Fig. 5.19 for an axi-symmetric Tokamak geometry. This implies that the equilibrium solution has to be mapped to the coordinate system chosen in the stability program [21]. Fourier analysis is performed in the two angular directions. Due to the periodicity of the equilibrium solution, Fourier terms in the toroidal direction decouple. For instance,

Fig. 5.19 The coordinates for
an axisymmetric Tokamak
geometry. s numbers the
magnetic flux surfaces, and Θ
and Φ are the poloidal and
toroidal angles, respectively

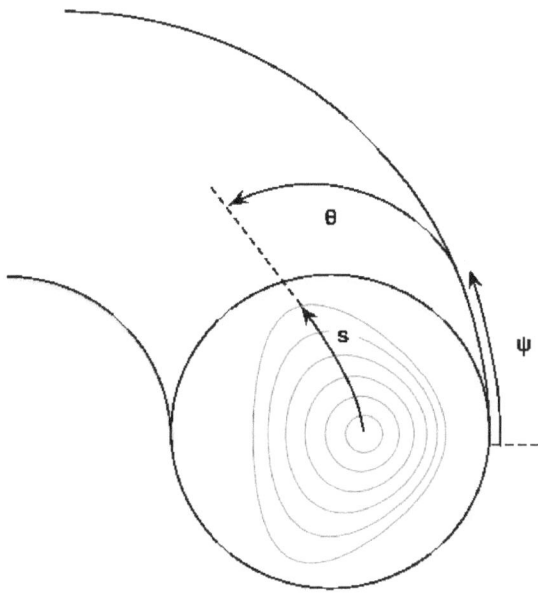

for the Wendelstein 7X geometry, Fig. 5.17, the number of periods in toroidal direction is equal to 5. Then, in the linear stability analysis, the following toroidal classes of modes exist (j is integer):

$$5j, -\infty < j < +\infty \qquad (5.7)$$
$$5j + 1, 5j - 1, -\infty < j < +\infty$$
$$5j + 2, 5j - 2, -\infty < j < +\infty$$

In poloidal direction, all the mode numbers couple. Special types of finite elements [61, 66] are applied in the flux coordinate direction that can satisfy internal constraints inherent in the ideal linear MHD equations. This approach delivers a high precision eigenvalue solution.

The coordinate mapping and the reconstruction of the equilibrium quantities take most of the computing time. The generalized eigenvalue problem is solved by an efficient inverse vector iteration process [5] that generally takes a small portion of the CPU time, but can become dominant for cases needing a big number of Fourier components. For the Wendelstein example, typically 16 poloidal and 8 toroidal modes, altogether 128, are chosen for each class of modes, Eq. 5.7. The tridiagonal matrix blocks per radial interval have then sizes of 512×512, and the matrix blocks overlap from one interval to the other by an 128×128 subblock. This subblock corresponds to the radial component of the variables that sit on the radial interval borders, whereas the toroidal and poloidal variables are at the center of each radial interval. This enables to first eliminate the internal variables, ending up with a block tridiagonal matrix for the radial vector component.

5.3.2 Optimization of VMEC

VMEC is one of the test examples to demonstrate optimization procedures. It has been optimized for the Cray and NEC vector machines. It admits that magnetic flux surfaces exist, and delivers as result their geometries. While VMEC is running at peak performance on a NEC SX-5, we concentrate our considerations on optimizing it for cache based RISC processors. The program is compiled with the Intel ifort Fortran compiler with the optimization options switched off. This is done to demonstrate the effect of the optimization steps.

The profiling, Fig. 5.18 indicates that the two subroutines $totzsp$ and $tomnsp$ demand 70% of the computing time. We concentrate the optimization process on subroutine tomnsp, Fig. 5.20, $totzsp$ has a very similar structure, and can be optimized accordingly. One realizes that the subroutine has been written such that the main memory accesses are all made with a stride=1. This seems to be optimal. The initial program sequence shown in Fig. 5.20 has 36 operations and 40 main memory accesses (all arrays with first index jk), leading to $V_a = \frac{36}{40}$. Switching the compiler optimization off, the total CPU time of the two subroutines is 1303 seconds and the total CPU time of VMEC is 1850 seconds.

In a first hand optimization step, Fig. 5.21, the long loop is cut into six smaller ones, keeping the same quantities in cache. The first two loops have a $V_a = \frac{10}{7}$ and

```
do 20 i = 1,ntheta2
 do 25 jk = 1,ns*nzeta
  temp1 = armn(jk,i,mparity)+xmpq(m,1)*arcon(jk,i,mparity)
  temp3 = azmn(jk,i,mparity)+xmpq(m,1)*azcon(jk,i,mparity)
  work2(jk,01) = work2(jk,01) + temp1*cosmui(i,m)
            + brmn(jk,i,mparity)*sinmum(i,m)
  work2(jk,02) = work2(jk,02) - crmn(jk,i,mparity)*cosmui(i,m)
  work2(jk,03) = work2(jk,03) + temp1*sinmu(i,m)
          + brmn(jk,i,mparity)*cosmumi(i,m)
  work2(jk,04) = work2(jk,04) - crmn(jk,i,mparity)*sinmu(i,m)
  work2(jk,05) = work2(jk,05) + temp3*cosmui(i,m)
          + bzmn(jk,i,mparity)*sinmum(i,m)
  work2(jk,06) = work2(jk,06) - czmn(jk,i,mparity)*cosmui(i,m)
  work2(jk,07) = work2(jk,07) + temp3*sinmu(i,m)
          + bzmn(jk,i,mparity)*cosmumi(i,m)
  work2(jk,08) = work2(jk,08) - czmn(jk,i,mparity)*sinmu(i,m)
 25       continue
 do 30 jk = 1,ns*nzeta
  work2(jk,09) = work2(jk,09) + blmn(jk,i,mparity)*sinmum(i,m)
  work2(jk,10) = work2(jk,10) - clmn(jk,i,mparity)*cosmui(i,m)
  work2(jk,11) = work2(jk,11) + blmn(jk,i,mparity)*cosmumi(i,m)
  work2(jk,12) = work2(jk,12) - clmn(jk,i,mparity)*sinmu(i,m)
 30       continue
 20       continue
```

Fig. 5.20 VMEC: Subroutine tomnsp. $Time(tomnsp) = 600s$, $Time(totzsp) = 700s$. Pipelining with stride = 1. $V_a = \frac{36}{40}$

```
do 20 i = 1,ntheta2
 do jk = 1,ns*nzeta
  temp1 = armn(jk,i,mparity)+xmpq(m,1)*arcon(jk,i,mparity)
  work2(jk,01) = work2(jk,01) + temp1*cosmui(i,m)
               + brmn(jk,i,mparity)*sinmum(i,m)
  work2(jk,03) = work2(jk,03) + temp1*sinmu(i,m)
               + brmn(jk,i,mparity)*cosmumi(i,m)
 enddo
 do jk = 1,ns*nzeta
  temp3 = azmn(jk,i,mparity)+xmpq(m,1)*azcon(jk,i,mparity)
  work2(jk,05) = work2(jk,05) + temp3*cosmui(i,m)
               + bzmn(jk,i,mparity)*sinmum(i,m)
  work2(jk,07) = work2(jk,07) + temp3*sinmu(i,m)
               + bzmn(jk,i,mparity)*cosmumi(i,m)
 enddo
 do jk = 1,ns*nzeta
  work2(jk,02) = work2(jk,02) - crmn(jk,i,mparity)*cosmui(i,m)
  work2(jk,04) = work2(jk,04) - crmn(jk,i,mparity)*sinmu(i,m)
 enddo
 do jk = 1,ns*nzeta
  work2(jk,06) = work2(jk,06) - czmn(jk,i,mparity)*cosmui(i,m)
  work2(jk,08) = work2(jk,08) - czmn(jk,i,mparity)*sinmu(i,m)
 enddo
 do jk = 1,ns*nzeta
  work2(jk,09) = work2(jk,09) + blmn(jk,i,mparity)*sinmum(i,m)
  work2(jk,11) = work2(jk,11) + blmn(jk,i,mparity)*cosmumi(i,m)
 enddo
 do jk = 1,ns*nzeta
  work2(jk,10) = work2(jk,10) - clmn(jk,i,mparity)*cosmui(i,m)
  work2(jk,12) = work2(jk,12) - clmn(jk,i,mparity)*sinmu(i,m)
 enddo
20 enddo
```

Fig. 5.21 VMEC: Subroutine `tomnsp`. First optimization step : smaller do-loops. $Time(tomnsp) = 400s$, $Time(totzsp) = 450s$. $V_a = \frac{10}{7}$ for the first two loops, $V_a = \frac{4}{5}$ for the following 4 loops, in average $V_a = \frac{36}{34}$

the following four loops have $V_a = \frac{4}{5}$, and in average $V_a = \frac{36}{34}$. With these smaller loops, the CPU time for the subroutine tomnsp went slightly down from initially 700 to 600 seconds, corresponding exactly to the ratio $\frac{40}{34}$ between the two V_a values.

In a second optimization step, Fig. 5.22, the most outer loop is moved in front of each of the six partial loops. The compiler now keeps the vectors $work2$ in cache for all j, $V_a = \frac{10}{3}$ for the first two loops, and $V_a = 4$ for the following four loops, and $V_a = \frac{36}{10}$ in average. As a consequence, the CPU time for tomnsp reduces to 170 seconds, corresponding to an additional CPU time reduction by a factor of 600/170 = 34/10, the latter ratio coming from the relative V_a ratio, showing that prediction is possible. The CPU time of the two dominant subroutines have gone down to about one third of the initial time, and the total CPU time of VMEC went down from 1850 to 1010 seconds, see Fig. 5.23.

```
do i = 1,ntheta2
 do jk = 1,ns*nzeta
  temp1 = armn(jk,i,mparity)+xmpq(m,1)*arcon(jk,i,mparity)
  work2(jk,01) = work2(jk,01) + temp1*cosmui(i,m)
              + brmn(jk,i,mparity)*sinmum(i,m)
  work2(jk,03) = work2(jk,03) + temp1*sinmu(i,m)
              + brmn(jk,i,mparity)*cosmumi(i,m)
 enddo
enddo

do i = 1,ntheta2
 do jk = 1,ns*nzeta
  temp3 = azmn(jk,i,mparity)+xmpq(m,1)*azcon(jk,i,mparity)
  work2(jk,05) = work2(jk,05) + temp3*cosmui(i,m)
              + bzmn(jk,i,mparity)*sinmum(i,m)
  work2(jk,07) = work2(jk,07) + temp3*sinmu(i,m)
              + bzmn(jk,i,mparity)*cosmumi(i,m)
 enddo
enddo

do i = 1,ntheta2
 do jk = 1,ns*nzeta
  work2(jk,02) = work2(jk,02) - crmn(jk,i,mparity)*cosmui(i,m)
   work2(jk,04) = work2(jk,04) - crmn(jk,i,mparity)*sinmu(i,m)
 enddo
enddo

do i = 1,ntheta2
 do jk = 1,ns*nzeta
  work2(jk,06) = work2(jk,06) - czmn(jk,i,mparity)*cosmui(i,m)
  work2(jk,08) = work2(jk,08) - czmn(jk,i,mparity)*sinmu(i,m)
 enddo
enddo

do i = 1,ntheta2
 do jk = 1,ns*nzeta
  work2(jk,09) = work2(jk,09) + blmn(jk,i,mparity)*sinmum(i,m)
  work2(jk,11) = work2(jk,11) + blmn(jk,i,mparity)*cosmumi(i,m)
 enddo
enddo

do i = 1,ntheta2
 do jk = 1,ns*nzeta
  work2(jk,10) = work2(jk,10) - clmn(jk,i,mparity)*cosmui(i,m)
  work2(jk,12) = work2(jk,12) - clmn(jk,i,mparity)*sinmu(i,m)
 enddo
enddo
```

Fig. 5.22 VMEC: Subroutine `tomnsp`. Second optimization step : reduce memory access through caching of *work*. $Time(tomnsp) = 170s$, $Time(totzsp) = 300s$. $V_a = \frac{10}{3}$ for the first two loops, $V_a = 4$ for the following four loops, in average $V_a = \frac{36}{10}$

```
%cum    self        self tot    times
29.43   300.07      300.07      3130   0.10 0.10 totzsp_.40
16.80   471.40      171.34      3124   0.05 0.05 totmnsp_.40
15.94   633.91      162.50      3124   0.05 0.05 forces_
12.80   764.45      130.54      3125   0.04 0.04 bcovar_
 7.17   837.56       73.11      3125   0.02 0.02 getiota_
 4.34   881.83       44.27      3128   0.01 0.01 alias_
 4.24   925.07       43.24      3130   0.01 0.01 jacobian_
 3.51   960.90       35.83      3130   0.01 0.32 funct3d_
 1.50   976.15       15.25    339277   0.00 0.00 sdot_
 1.46   991.04       14.89       288   0.05 0.05 precondn_
```

Fig. 5.23 Profiling of VMEC after optimization of `totzsp` and `tomnsp`

If the vector length $ns * nzeta$ is too big to be kept in cache during the i loop, it is advised to perform the jk loop in chunks for which the work2 arrays can remain in cache.

We have to mention here that when using the highest optimization option of the newest compilers, a very similar speedup can be reached automatically. Why should core optimization be done by hand? In general not, but if loops become too complicated, it can happen that a compiler is not further able to optimize efficiently, and simplifying coding can support the compiler to make a better job.

5.3.3 Optimization of TERPSICHORE

5.3.3.1 The world's fastest program

The plasma physics code TERPSICHORE [5] was the fastest program in 1989, winning the Cray Gigaflop Performance Award [6] with a sustained performance of 1.708 GF/s on an eight processor Cray Y-MP that had a peak performance at 2.496 GF/s (see Table 5.2). To reach those 68.4% of peak, the initial code was parallelized using what was called Microtasking, a semi-automatic compiler option available on the Cray Y-MP. The multiple loops were distributed among the processors through directives interpreted by the compiler. The slightly modified and still portable program reached a speedup of 7.35 out of 8, demanding an increase of the number of operations by 15.2%, and reducing the real achieved efficiency to 58%.

Table 5.2 Benchmark measurements of TERPSICHORE. The efficiencies for the IBM p575 and the Intel Woodcrest were best measurements achieved with highest optimization options, but without hand optimization, and no parallelization

Year	Machine	Cores	R_∞ [GF/s]	R_a [GF/s]	g_p
1989	Cray Y-MP	8	2.496	1.708	0.58
2003	NEC-SX5	1	8	4.07	0.51
2007	IBM p575	1	6	1.176	0.19
2007	Intel Woodcrest	4	21.33	1.27	0.06

Thanks to the help by a highly skilled Cray engineer, the parallelization work took just 3 weeks of time.

5.3.3.2 The NEC version

The Cray Y-MP version was ported to the NEC SX-3 at CSCS, and ran on the SX series until 2007. In the NEC version the Cray-related library modules of the eigenvalue solver were replaced by Fortran subroutines. The highly optimized NEC Fortran compiler was able to produce an executable of similar efficiency than on the Y-MP. On the NEC SX-5 4.07 GF/s were achieved, corresponding to 51% of peak.

5.3.3.3 The single core RISC version

The TERPSICHORE was first ported from a NEC SX-5 onto an IBM p575 cluster with 6 GF/s peak per core. Only 1.094 GF/s were measured for a case needing 14 GB of main memory, and 1.176 GF/s (19.6% of peak) after an inlining optimization step. A profile has been made on the IBM machine (Table 5.3). Two subroutines take about 60% of the total CPU time, i.e. the matrix times matrix multiplication with 41%, and the matrix solver with 19%.

To find out if improvements could be done, TERPSICHORE was sent to one of the authors. He first replaced the NEC written basic algebra modules by the highly optimized LAPACK [7] modules DGEMM (for the matrix times matrix operation) and DGESV (matrix solver). The result was close to a factor of 2 reduction of the CPU time. MXM and FMIND then were not the dominant subroutines any more.

The profiling of the stability program reveals that the subroutines FOURIN and LAMNEW take a good portion of the computing time. We shall concentrate on those modules. To study optimization by hand, the lowest optimization option $-O1$ was used again for compiling.

The FOURIN program was optimized for the NEC SX series, Fig. 5.24. The innermost jk loop is much longer than the external lr loop. For the vector machine, the longest loop should be the innermost one, and stride odd, which is the case. In RISC machines, the jk loop leads to main memory conflicts due to non stride 1 accesses of the arrays tsc and tss, Fig. 5.24.

Table 5.3 TERPSICHORE profiling on the IBM p575

Name	sec	%	
MXM	2183	40.76	Matrix * Matrix multiply
FMIND	1008	18.82	Matrix solve
FOURIN	400	7.48	Fourier integrals
LHSMAT	275	5.13	Left hand side matrix construction
LAMNEW	207	3.86	Renormalization
ENERGY	165	3.07	Diagnose energy
RHSMAT	154	2.87	Right hand side matrix construction
VMTOBO	149	2.78	Transformation to Boozer coordinates
METRIC	139	2.60	Metric elements in Boozer frame
Total	5356	100	

```
do I = NPR,NIV,NPROCS

    do lr = 1,lss
        do jk = 1,njk
            cna(1,lr) = cna(1,lr)+gppl(jk,i)*tsc(lr,jk)
            cna(2,lr) = cna(2,lr)+gtpl(jk,i)*tsc(lr,jk)
            cna(3,lr) = cna(3,lr)+gttl(jk,i)*tsc(lr,jk)
        enddo
    enddo

    do lr = 1,lss
        do jk = 1,njk
            cnb(1,lr) = cnb(1,lr)+gstl(jk,i)*tss(lr,jk)
            cnb(2,lr) = cnb(2,lr)+gssl(jk,i)*tsc(lr,jk)
            cnb(3,lr) = cnb(3,lr)+gssu(jk,i)*tss(lr,jk)
        enddo
    enddo

enddo
```

Fig. 5.24 TERPSICHORE: Subroutine FOURIN. Before optimization. Stride is not 1 for tss and tsc and that loop takes **55 seconds**

When the jk and lr loops are interchanged, Fig. 5.25, the arrays tsc and tss are accessed with stride 1. The quantities cna and cnb are kept in cache by the compiler. The 55 seconds CPU time of the initial version of FOURIN has been reduced to 5 seconds, a nice 11 fold acceleration.

```
do I = NPR,NIV,NPROCS

    do jk = 1,njk
        do lr = 1,lss
            cna(1,lr) = cna(1,lr)+gppl(jk,i)*tsc(lr,jk)
            cna(2,lr) = cna(2,lr)+gtpl(jk,i)*tsc(lr,jk)
            cna(3,lr) = cna(3,lr)+gttl(jk,i)*tsc(lr,jk)
        enddo
    enddo

    do jk = 1,njk
        do lr = 1,lss
            cnb(1,lr) = cnb(1,lr)+gstl(jk,i)*tss(lr,jk)
            cnb(2,lr) = cnb(2,lr)+gssl(jk,i)*tsc(lr,jk)
            cnb(3,lr) = cnb(3,lr)+gssu(jk,i)*tss(lr,jk)
        enddo
    enddo

enddo
```

Fig. 5.25 TERPSICHORE: Subroutine FOURIN. After optimization. Indexes have been exchanged. The first inner loop has a $V_a = 6$ and the second inner loop has a $V_a = 3$. The stride is 1 and lss is small enough to keep cna and cnb in cache. The loop takes now **5 seconds**

```
do jk = 1,njk
    do ls = 1,lss
        gla(1,ls) = gla(1,ls)+pgppv(jk)*tsc(ls,jk)
        gla(2,ls) = gla(2,ls)+pgtpv(jk)*tsc(ls,jk)
        gla(3,ls) = gla(3,ls)+pgttv(jk)*tsc(ls,jk)
    enddo
enddo
```

Fig. 5.26 TERPSICHORE: Subroutine LAMNEW (called N1 times). Before optimization. That routine takes **87 seconds**

```
do ls = 1,lss
    do jk = 1,njk
        gla(ls,1) = gla(ls,1)+pgppv(jk)*tsc(ls,jk)
        gla(ls,2) = gla(ls,2)+pgtpv(jk)*tsc(ls,jk)
        gla(ls,3) = gla(ls,3)+pgttv(jk)*tsc(ls,jk)
    enddo
enddo
```

Fig. 5.27 TERPSICHORE: Subroutine LAMNEW (called N1 times). After optimization (the indexes have been exchanged on gla). That routine takes now **12 seconds**

In the LAMNEW subroutine, Fig. 5.26, the *gla* array showed a stride 1 problem. In fact, this array should have been kept in cache, but was not. By exchanging the indexes, the *gla* array is accessed with stride 1, and the CPU time moved from 87 seconds to 12 seconds, Fig. 5.27.

5.3.4 Conclusions for single core optimization

We learned with this example that porting a code to other types of computer architectures can lead to a substantial performance reduction. Reaching high performance requires an adaptation of the application to the underlying hardware. Increasing efficiency of a code by optimizing it reduces the waiting time of the research scientist is reduced, her/his productivity increased, and energy consumption reduced. However, the most recent compilers reach a very high level of automatic optimization. Interventions in the code are now rarely needed.

Chapter 6
Node optimization

L'avenir n'est pas une amélioration du présent. C'est autre chose.
Elsa Triolet, French writer (1896 – 1970)

Abstract Once the code optimization is done at a single core-level (chapter 5), the adaptation to a multi-core architectures can be addressed. Different shared memory machines (SMP) and non uniform memory access (NUMA) architectures are presented. Vector architectures are briefly discussed to show how memory access bounded applications can benefit from such machines. Examples in CFD and plasma physics are presented.

6.1 Shared memory computer architectures

6.1.1 SMP/NUMA architectures

SMP (Symmetric Memory Processor) is a parallel computer architecture with a shared main memory in which all the processors see the memory in a totally symmetric manner. In the case of a NUMA (Non-Uniform Memory Access) machine, the memory is distributed among all the processors. The system considers it as a virtual shared memory that can, from the user point of view, be accessed as in a SMP machine. If the memory of another processor is accessed, the latency increases, and the bandwidth decreases. As a consequence, the computing time in a NUMA architecture cannot be predicted with precision.

In SMP and NUMA architectures the parallelism can be expressed with the OpenMP library, with threads, or with the more general MPI or PVM libraries. OpenMP and threads are applied to shared main memory, thus, do not need data parallelism, whereas MPI and PVM do, demanding more complex message passing operations.

6.1.1.1 SMP server architecture

A typical example of a SMP machine is the Intel Xeon (Fig. 6.1) server including 2 cores. The motherboard of this machine connects to both cores, and has one bus

R. Gruber, V. Keller, *HPC@Green IT*,
DOI 10.1007/978-3-642-01789-6_6, © Springer-Verlag Berlin Heidelberg 2010

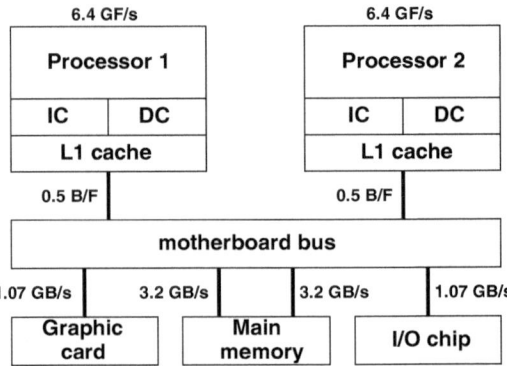

Fig. 6.1 The Intel Xeon server with 2 processors. The main memory is accessed through the motherboard. The memory access path is the same as for a single processor architecture, Fig. 5.6. The V_m parameter doubles

to access the main memory as in a single processor PC. This implies that the main local memory bandwidth per processor is now half, and V_m doubles. For applications dominated by the memory bandwidth, doubling the number of processors without increasing the memory bandwidth does not change the overall performance, sometimes the overall performance can even be slightly reduced.

6.1.1.2 SMP vector architecture

Vector machines have been the top supercomputers in the last century, Fig. 2.4, outperforming desktop computers by up to 3 orders of magnitude. With the RISC architectures arising in PCs and laptops this performance gap has been reduced and the most famous Cray vector supercomputers disappeared on the market. The NEC SX series are the only vector machines still manufactured, the newest one being the NEC SX-9. Schematically, the architecture of an older NEC SX-5 vector processor is shown in Figs. 6.2 and 6.3. The functional units access the data from the 8 vector registers (VR) and write the results back to a data register (DR) from which the data can flow into another VR to be directly reaccessed to execute another vector operation. The memory in such a vector machine is accessed through a fully scalable memory switch with a memory bandwidth characterized by $V_m = 1$ (one operand per result) for a NEC SX-5, and $V_m = 2$ for newer machines in the SX Series. The memory switch of an 8 processor NEC SX-5 has $16'384=2^{14}$ links to the $16'384$ memory banks, $2'048$ banks per processor.

To show how main memory access functions in vector machine, a Fortran array $x(3, 64)$ with 192 components is written. The first component x(1,1), is stored in bank number 1, the second component $x(2, 1)$ is placed in bank number 2, then $x(3, 1)$ in bank 3, and $x(1, 2)$ in bank number 4. If one reads the first line, i.e. $x(1, 1)$ (component 1), $x(1, 2)$ (component 4), ..,$x(1, 64)$ (component 190), The components are stored with stride=3. The component $x(1, 1)$ is in bank number 1, $x(1, 2)$ in bank 4, the component number 64, i.e. $x(1, 22)$, is stored in bank 64.

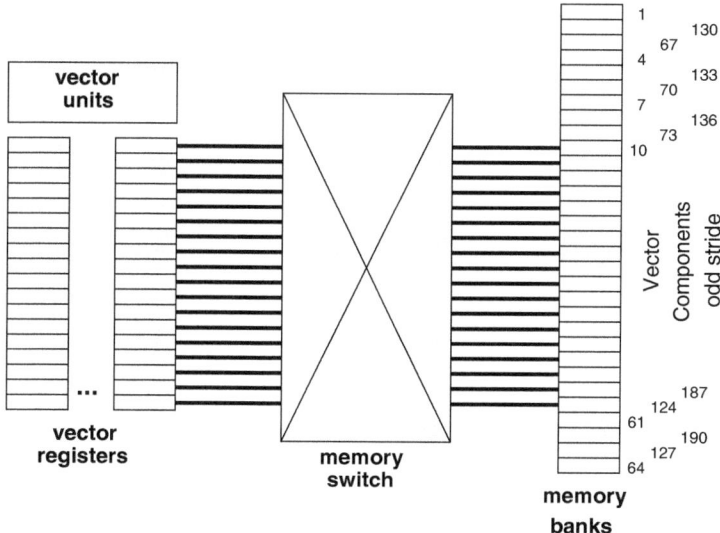

Fig. 6.2 Schema of a vector machine with a main memory subsystem of 64 banks. The vector functional units access the vector registers. Vector registers are filled through the memory switch that accesses the memory banks of the memory shared among different vector units, each one having add and multiply pipelines. On the right hand side is indicated how a vector of length 64 with stride 3 is stored in a memory that is subdivided into 64 banks. The 64 components are distributed among all the 64 memory banks

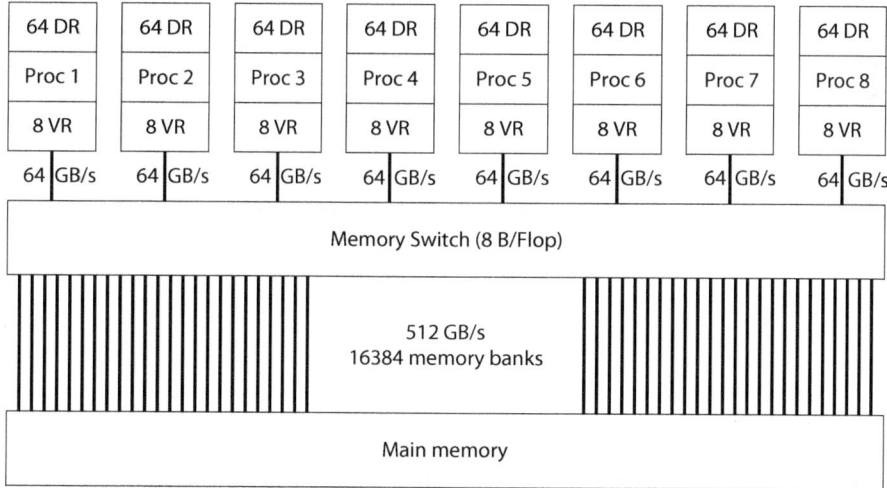

Fig. 6.3 The NEC SX-5 architecture with 8 processors. Altogether 16'384 banks are needed to feed the 8 processors with 8 add and 8 multiply pipelines each. The overall memory bandwidth is 512 GB/s, the peak performance 64 GF/s, and $V_m = 1$. Besides the vector units there is also a superscalar processor that takes care of the scalar operations

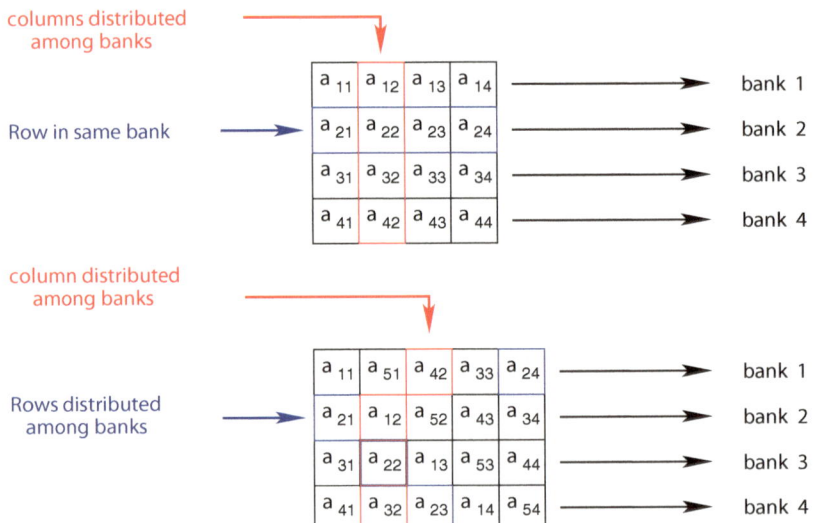

Fig. 6.4 Memory conflict if a row of a 4x4 matrix is read in a Fortran program (upper situation). In a vector machine memory conflict can be resolved if the matrix A is dimensioned with $a(5, 4)$ instead of $a(4, 4)$. Then $a(1, 2)$ gets to bank 2 instead of 1

Since storage is made in round robin manner, component number 67, i.e. $x(1, 23)$, is in bank number 3, the component number 127, $x(1, 33)$, is in bank 63. This situation is depicted in Fig. 6.4. One realizes that after having read all the 64 components of the first line all the 64 banks have been visited exactly one time. Now, we can read line 2, without any main memory conflict. This is different from a RISC processor. There, top bandwidth is obtained when main memory writes and reads with stride=1 are performed.

Can we get main memory conflicts in vector machines? Suppose that we have a matrix declared $a(64, 64)$. When one reads a line in Fortran, the $a(1, 1)$ is component 1 in bank 1, $a(1, 2)$ is component 65 in bank 1, $a(1, 3)$ is component 129 in bank 1, etc. The whole matrix line is in bank 1, and the main memory bandwidth drops by a factor of 64. One would recover this loss factor if one would declare $a(65, 64)$. The loss through main memory conflicts in vector machines can become much larger than in PCs or laptops.

Why do we discuss old stuff? First, there is still the NEC SX-9 supercomputer on the market. Then, we believe that future main memory subsystems will come back to the banking mechanism to guarantee a high bandwidth for upcoming many core nodes.

The latency time in a vector machine is directly related to the quantity n_{12} is $n_{\frac{1}{2}}$ [78] that measures the vector length needed to achieve half of the peak performance. In fact, this quantity measures the latency time, is directly related to the number of main memory banks in the system, and varies for different operations. In a NEC SX-5 for instance $n_{\frac{1}{2}}$ is small when the vector registers can be used to store intermediate results, as for DGEMM, but if data has to be brought from main memory, as for SMXV, $n_{\frac{1}{2}}$ can directly be related to the number of banks. This means that a vector machine often achieves high efficiency if the vector length is huge. All 3D finite element or finite volume applications with millions of mesh points reach high performance on vector machines.

Main memory systems in PCs or laptops are based on the "paging" mechanism. When reading a byte or a 8B word from main memory, 64B stored with stride=1 are transferred. This makes it clear that data should best be read in chunks of 64B and with stride=1. Then, pipelining in the data transfer from main memory to highest level cache can be done in a most efficient manner.

6.1.1.3 The NUMA architecture

The AMD Opteron machine shown in Fig. 6.5 is an example of a NUMA architecture. The two processors have their own main memory subsystem, the operating system considers them as to be shared. If data is positioned on the local processor the execution is faster than if data sits on the other processor.

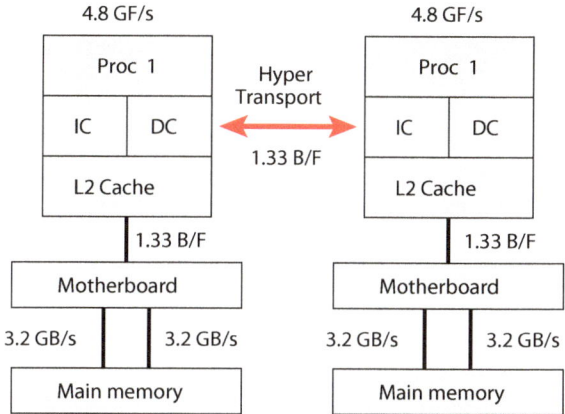

Fig. 6.5 The Opteron NUMA server. Each processor has its own local main memory. The system considers the physically distributed memory as a (virtually) shared memory. It places the data in the memory in a more or less statistical manner. If data is sitting on the local memory, performance is bigger than when it sits on the other processor. The HyperTransport should guarantee a close to perfect scalability

6.1.2 The Cell

IBM has developed with Sony and Toshiba the so-called Cell chip (see Fig. 6.6) including one control Power 4 core with a main memory up to 4GB and 8 coprocessors connected to the control core by a 200 GB/s high speed bus.

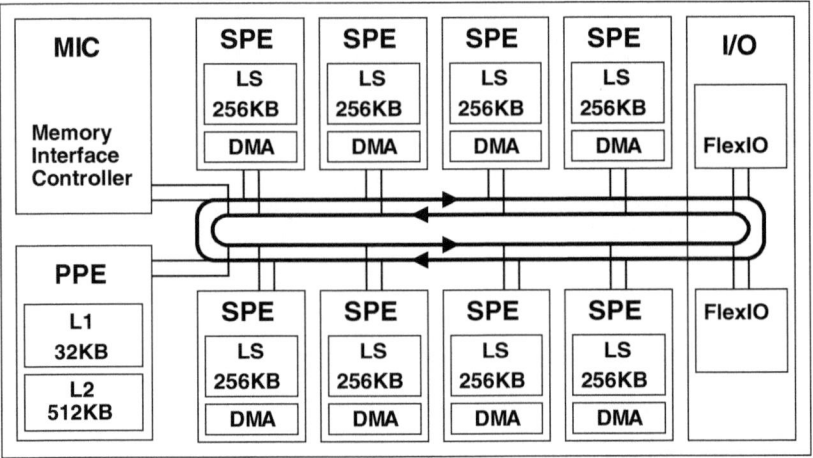

Fig. 6.6 The IBM Cell node

6.1.3 GPGPU for HPC

The use of graphics processing units or GPUs for HPC is becoming an accepted alternative to CPU-based systems. A potentially large number of parallel threads, fast memory access, a good performance/Watt ratio, and outstanding speedup factors in comparison with standard computers render them attractive in many scientific computing disciplines. In 2007, NVIDIA introduced the Compute Unified Device Architecture (CUDA), an extended C programming model, offering high-level access to NVIDIA's Tesla architecture.

6.1.3.1 The Tesla architecture

The Tesla architecture [98] is based on a scalable processor array. Figure 6.7 shows a block diagram of the GeForce GTX 280 GPU with 240 Streaming-Processor (SP) cores organized in ten independent processing units called Texture/Processor Clusters (TPCs). Each TPC Figure 6.8 has one texture unit, three sets of processing units called Streaming Multiprocessors (SM), Fig. 6.12, an SM Controller (SMC) and a Geometry Controller. An SM is an array of Streaming Processor (SP) cores, eight to be specific, along with two more processors called Special Function Units (SFUs). Each SFU has four FP multiply units which are used for transcendental operations (e.g. sin, cos) and interpolation. In addition to the processor cores in an SM, there's a very small instruction cache, 16K 32-bit registers, 16-Kbyte read/write shared memory, and supports up 1024 co-resident threads. The texture block includes texture addressing and filtering logic as well as an L1 texture cache.

The result is a GPU with a total possible population of 30K threads that delivers 933 GF/s ($= 1.296$ GHz per core $\times$ 3 ops/Hz $\times$ 240 cores) at peak. Double precision arithmetic runs 12 times smaller, i.e. up to 78 GF/s.

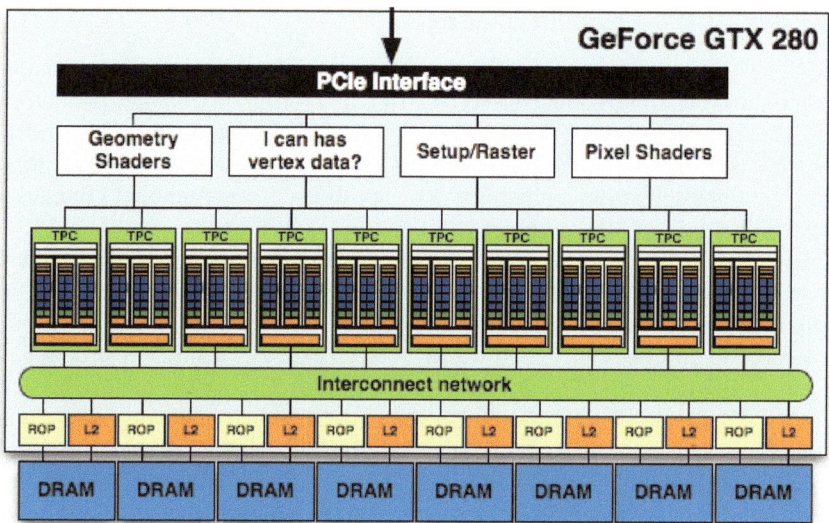

Fig. 6.7 The NVIDIA GeForce GTX 280

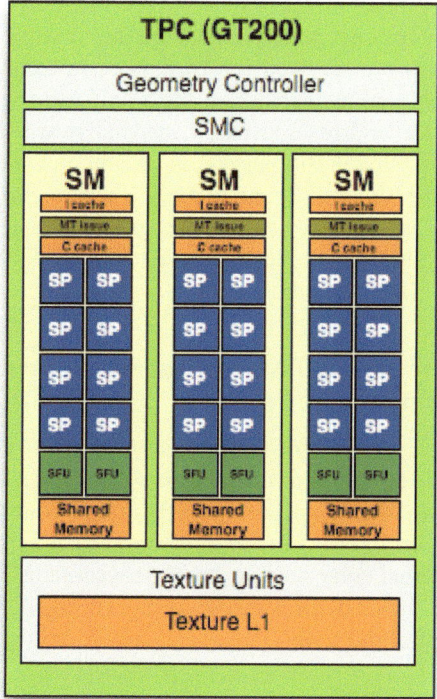

Fig. 6.8 The texture/processor cluster

6.1.3.2 The CUDA computation model

The CUDA computation model offers the GPU as a massively multi-threaded architecture to the CPU. In this context a serial host (CPU) program executes parallel programs known as kernels on a device (GPU). A kernel executed by a large number of threads is a SPMD (Single Program Multiple Data) model meaning that each thread execute the same scalar sequential program. The user-defined number of threads are organized into a grid of thread blocks [116].

Thread creation, scheduling, and management is performed entirely in hardware. To manage this large population of threads efficiently, the GPU employs a SIMT (Single Instruction Multiple Threads) architecture in which the threads of a block are executed in groups of 32 called warps. A warp executes a single instruction at a time across all its threads. The threads of a warp are free to follow their own execution path, and all such execution divergence is handled automatically in hardware. However, it is substantially more efficient for threads to follow the same execution path for the bulk of the computation.

The threads of a warp are also free to use arbitrary addresses when accessing off-chip memory with LOAD/STORE operations. Accessing scattered locations results in memory divergence and requires the processor to perform one memory transaction per thread. On the other hand, if the locations being accessed are sufficiently close together, the per-thread operations can be coalesced for greater memory efficiency. Global memory is conceptually organized into a sequence of 128-byte segments. Memory requests are serviced for 16 threads (a half-warp) at a time. The number of memory transactions performed for a half-warp will be the number of segments touched by the addresses used by that half-warp. If a memory request performed by a half-warp touches precisely one segment, we call this request fully coalesced, and one in which each thread touches a separate segment we call uncoalesced. If only the upper or lower half of a segment is accessed, the size of the transaction is reduced said two research scientist within NVIDIA Research [15]. The Fig. 6.9 illustrates SIMT scheduling.

6.1.3.3 Basic structure of a CUDA program: simple example of matrix*vector product

Let us consider a simple example of matrix-vector product [123], Figs. 6.10, 6.11, for illustrating the basic structure of a CUDA program code. Every thread uploads a couple of elements to shared memory, every thread computes the dot product between one line of matrix M and vector x. The _global_qualifier indicates that the function is a kernel entry point, blockDim contains the block dimensions, blockIdx the block index within the grid and threadIdx the thread index within the block. A call of the form

```
vecMat<<<blockGrid,threadBlock,w*sizeof(float)>>>
```

is a parallel kernel invocation that will launch blockGrid thread blocks of threadBlock threads each, w*sizeof(float) gives the number of bytes

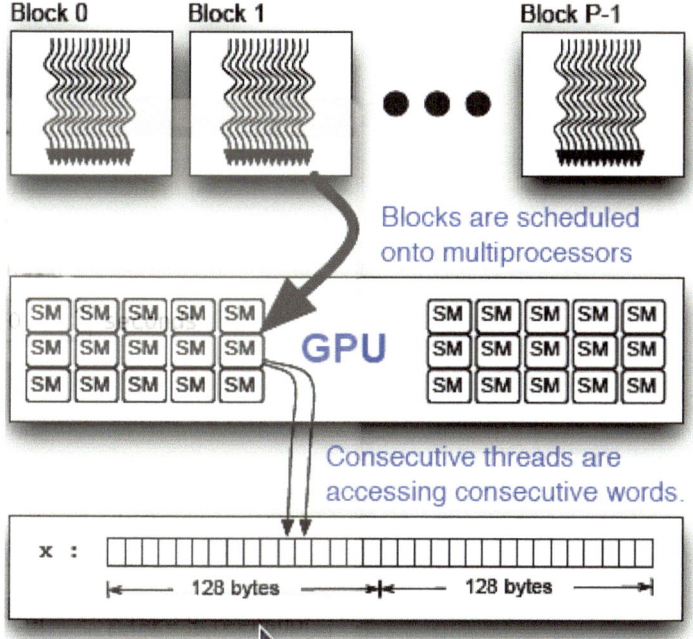

Fig. 6.9 The SIMT scheduling

dynamically allocated in shared memory. CUDA threads may access data from multiple memory spaces:

- Register dedicated HW - single cycle
- Shared Memory dedicated HW - single cycle - shared between threads in the same block
- Global Memory DRAM, no cache - slow
- Constant Memory DRAM, cached, 1...10s...100s of cycles - depends on cache locality
- Texture Memory DRAM, cached, 1...10s...100s of cycles - depends on cache locality
- Instruction Memory (invisible) DRAM, cached

Synchronization for all threads in CUDA is done implicitly through function _syncthreads(). This function will coordinate communication among threads of the same block (see Fig. 6.10). N. Fujimoto [54] and Nathan Bell et al. [15] have demonstrated how to achieve significant percentages of peak performance (floating point and bandwidth) on dense and sparse matrix on CUDA. If one considers the diagonal format representation (formed by two arrays: data, which stores the nonzero values, and offsets, which stores the offset of each diagonal from the main diagonal) performance of 20 GF/s (device: GPU GTX 280, Matrix Laplace 9pt, size: 1,000,000 × 1000,000, precision: single precision) can be obtained.

```
// upload M and x
cudaMemcpy( gpuMat, hostMat,
    w*h * sizeof(float), cudaMemcpyHostToDevice);
cudaMemcpy( gpuVec, hostVec,
    w * sizeof(float), cudaMemcpyHostToDevice);

// compute the block and grid dimensions
dim3 threadBlock( MAX_THREADS, 1 );
dim3 blockGrid( h / MAX_THREADS
+ 1, 1, 1);
 vecMat<<< blockGrid, threadBlock, w * sizeof(float) >>>
( gpuResVec,gpuMat, gpuVec, w,h, w / MAX_THREADS);

// forces runtime to wait until all
// preceding device tasks have finished
cudaThreadSynchronize();

// download result y
cudaMemcpy( hostResVec, gpuResVec,
    h * sizeof(float), cudaMemcpyDeviceToHost);


_global_void vecMat(float *_dst, const float* _mat,
    const float* _v, int _w, int _h, int nIter )
{
    extern _shared_ float vec[];
    int i = blockIdx.x * blockDim.x + threadIdx.x;
    float res = 0.; int vOffs = 0;

    // load x into shared memory
    for (int iter = 0; iter < nIter;
        ++iter, vOffs += blockDim.x) {
        vec[vOffs + threadIdx.x] = _v[vOffs + threadIdx.x];
    }
    // make sure all threads have written their parts
    _syncthreads();

    // now compute the dot product again
    // use elements of x loaded by other threads!
    if (i < _h) {
        for (int j = 0; j < _w; ++j) {
            res += _mat[offs + j ] * vec[j];
        }
        _dst[i] = res;
    }
}
```

Fig. 6.10 Sparse matrix*vector multiplication (SpMV) with CUDA N. and M. Garland

```
_global_ void spmv_dia_kernel ( const int num_rows ,
    const int num_cols ,
    const int num_diags ,
    const int * offsets ,
    const float * data ,
    const float * x,
    float * y)
{
    int row = blockDim.x * blockIdx.x + threadIdx.x ;
    if(row < num_rows ){
        float dot = 0;

        for ( int n = 0; n < num_diags ; n ++){
            int col = row + offsets [n];
            float val = data [ num_rows * n + row ];

            if( col >= 0 \&\& col < num_cols )
                dot += val * x[col ];
        }
        y[ row ] += dot;
    }
}
```

Fig. 6.11 Example of parallelizing with CUDA. From [15]

In [15] is said "Parallelizing SpMV for the diagonal format is straightforward: one thread is assigned to each row of the matrix. In Fig. 6.11, each thread first computes a sequential thread index, row, and then computes the (sparse) dot product between the corresponding matrix row and the x vector... all SpMV kernels can benefit from the texture cache present on all CUDA-capable devices...accessing the x vector through the texture cache often improves performance considerably. In the cached variants of our kernels, reads of the form x[j] are replaced with a texture fetch instruction tex1Dfetch(x tex, j), leaving the rest of the kernel unchanged."

6.2 Node comparison and OpenMP

The Table 6.1 gives the parameters of RISC computers and the measured performances for two characteristic benchmarks: DGEMM (measuring peak performance), and SMXV (measuring peak main memory bandwidth). We remind that N_{CPU} is the number of processors, each processor has its own main memory and its own lowest level cache, and N_{core} is the number of cores per processors. The total number of cores is $N_{CPU}*N_{core}$. Other machine parameters are the frequency F, the peak double precision performance R_∞ of the active cores, the peak main memory bandwidth M_∞ in units of 10^9 64bit words per second, the number of functional units ω_{core}, and $V_m = R_\infty/M_\infty$. R_∞^* is the peak double precision performance of the SSE hardware, R_a^{DGEMM} is the measured performance of the ScaLA-PACK kernel DGEMM that is optimally implemented on the SSE hardware, and

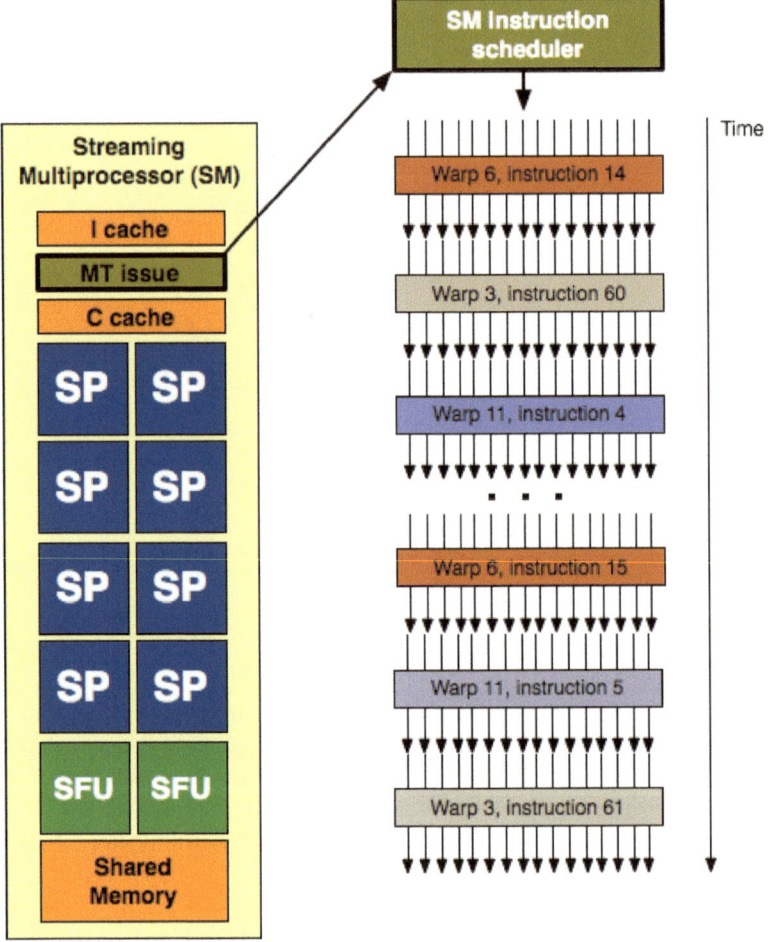

Fig. 6.12 The streeming multiprocessor

$g_p^* = R_a^{DGEMM}/R_\infty^*$ is the efficiency of the SSE hardware for this benchmark. The last four parameters are measures of the main memory subsystem. R_a^{SMXV} is the performance measured with the sparse matrix times vector kernel SMXV, $g_m = R_a^{SMXV}/M_\infty$, $V_m = R_a^{SMXV}/R_\infty$, and $V_m^* = R_a^{SMXV}/R_a^{DGEMM}$.

Two double precision benchmarks are presented: the Lapack library DGEMM subroutine measures the peak performance of the underlying hardware, which is relevant to the HPL benchmark, and the SMXV benchmark measures the real main memory bandwidth. The efficiencies $g_p^* = R_a^{DGEMM}/R_\infty^*$ of the processors are deduced from the DGEMM measurements, the efficiencies $g_m = M_a^{SMXV}/M_\infty$ of the main memory subsystems are deduced from the SMXV measurements. V_a being equal to 1, the main memory bandwidth M_a^{SMXV} in Gw/s is measured through $M_a^{SMXV} = R_a^{SMXV}$, where R_a^{SMXV} is measured in GF/s. We have to mention here

Table 6.1 Characteristics of a few RISC nodes on the market

Machine	N_{CPU}	N_{cores}	F GHz	R_∞ GF/s	M_∞ Gw/s	V_∞ F/w	R^*_∞ GF/s	R_a^{DGEMM} GF/s	g^*_p	R_a^{SMXV} GF/s	g_m	V_m F/w	V_m^{SSE} F/w
Pentium4 2003	1	1	2.8	5.6	0.8	7				0.4	0.5	14	
Itanium 2003	1	1	1.3	5.2	0.8	6.5				0.65	0.81	8	
2003	1	2	1.3	10.4	0.8	13	10.4			2x0.33	0.81	16	
Blue Gene/L 2004	1	1	0.7	2.8	0.33	4.2	2.8	2.14	0.76	0.10	0.3	14	21.4
	2	1	0.7	5.6	0.67	8.4	5.6	4.28	0.76	2x0.10	0.3	14	21.4
Woodcrest 2005	1	1	2.67	5.33	2.67	2	10.67	9.7	0.91	0.48	0.18	11	22
	2	1	2.67	10.67	2.67	4	21.33	18.6	0.87	2x0.4	0.30	12.5	25
	2	2	2.67	21.33	2.67	8	42.67	34.5	0.81	4x0.21	0.31	24	48
Harpertown 2007	1	1	2.33	4.67	2.67	1.75	9.33	8.6	0.92	0.47	0.18	9.7	19.4
	2	1	2.33	9.33	2.67	3.5	18.67	16.3	0.87	2x0.4	0.3	12	24
	2	2	2.33	18.67	2.67	7	37.33	31.4	0.84	4x0.22	0.33	21	42
	2	4	2.33	37.33	2.67	14	74.67	53.8	0.72	8x0.11	0.33	42	84
Nehalem 2009	1	1	2.7	5.4	8	0.67	10.8	10.7	0.99	1.4	0.18	3.7	7.5
	1	2	2.7	10.8	8	1.35	21.6	21.3	0.99	2x0.91	0.23	5.9	11.8
	1	4	2.7	21.6	8	2.7	43.2	41.0	0.95	4x0.50	0.25	10.8	21.6
Atom 2009	1	1	0.8	1.6	0.53	3.0				0.28	0.52	5.7	
	1	2	0.8	3.2	0.53	6.0				2x0.14	0.52	11.4	
GTX 280 2009	10	24	1.296	78	15	5.2		74	0.95	10	0.66		
	10	24	1.3	78						16			

that the DGEMM efficiencies are slightly higher than the HPL peaks. In fact, the HPL benchmark has two fundamental steps: the computationally dominant matrix decomposition that is based on subsequent DGEMM operations, and the backsolve based on DGEMV operations.

Let us try to understand the measurements in Table 6.1. When looking into the DGEMM measurements, all machines reach between 76% ($g_p^* = 0.76$) to 97% ($g_p^* = 0.97$) peak performance. In the Intel Xeon nodes (Woodcrest, Harpertown, and Nehalem), this peak is relative to the special SSE hardware that runs at the same frequency as the RISC processors, but four results per cycle period can be obtained instead of two. The DGEMM kernel has specially been assembler coded on this hardware to run at close to peak as in vector machines. Due to the Turbo-booster facility in the Nehalem node, the performance can even be higher than the theoretical peak, but the energy consumption is increased. In main memory bound applications, the Turbobooster does not contribute to a performance increase, and should be switched off.

In Table 6.2 are shown a few measurement of the matrix*matrix multiply algorithm written in Fortran, Fig. 6.13. In fact, these are the same triple loops, Fig. 5.7, with OpenMP directives to distribute the most upper loop to the cores. The results show almost ideal scalability with the number of cores, and 70 % efficiencies with respect to the DGEMM results in Table 6.1. The results with the kij loop, Fig. 6.14 are slightly larger than those for the kji loop at Fig. 6.13. Despite of this small difference, the results show that the new compilers are able to activate SSE kernels and to fully profit from OpenMP directives.

Let us now concentrate on the SMXV benchmarks. The R_a^{SMXV} performance measurements show efficiencies between 18% (g_m=0.18) for a one core Nehalem and 81% (g_m=0.81) for an Itanium processor. Note that in vector machines, $g_m > 0.9$.

Table 6.2 OpenMP measurements on a single processor quadri-core Nehalem

Machine	N_{CPU}	N_{cores}	F GHz	R_∞^{SSE} GF/s	R_a^{DGEMM} GF/s	R_a^{kji} GF/s	R_a^{kij} GF/s
Nehalem	1	1	2.7	10.8	10.7	7.84	7.87
2009	1	2	2.7	21.6	21.3	15.61	15.64
	1	4	2.7	43.2	41.0	28.47	28.96

```
c$OMP parallel do private(i,j,k)
for k = 1 to n do
  for j = 1 to n do
    for i = 1 to n do
      c(j, k) = c(j, k) + a(j, i) * b(i, k)
    end for
  end for
end for
c$OMP end parallel do
```

Fig. 6.13 Standard triple loop for a matrix times matrix with OpenMP directives to distribute k loop

```
c$OMP parallel do private(i,j,k)
for i = 1 to n do
    for k = 1 to n do
        for j = 1 to n do
            c(j, k) = c(j, k) + a(j, i) * b(i, k)
        end for
    end for
end for
c$OMP end parallel do
```

Fig. 6.14 Standard triple loop for a matrix times matrix with OpenMP directives to distribute i loop, leading to a race condition, and wrong results

The main memory subsystem of the Itanium nodes is highly efficient. There is a tendency towards lower g_m values when the number of cores increases. What can be seen in Table 6.1 is the stepwise increase in the main memory bandwidth: a factor of 3 from Pentium4 (M_∞=0.8 with 50% efficiency) to Woodcrest (M_∞=2.67 with 33% efficiency), a factor of 3 from Woodcrest to Nehalem (M_∞=8 with 25% efficiency). On the Woodcrest and the Harpertown nodes, the main memory bandwidths almost increase by a factor of 2 when going from one to 2 processors, but this total bandwidth is only very slightly increasing when the number of cores goes from 1 to 2, and there is no increase at all in the total main memory bandwidth when going from 2 to 4 cores in the Harpertown node.

In Table 6.1 is also added the new very low energy Intel node, called Atom that appeared also in 2009 as Nehalem. One realizes that the top main memory bandwidth is 15 times smaller than on a Nehalem, the really measured bandwidth is only 7 times smaller. If the number of cores per processor in a Nehalem machine increases to 6, the per core main memory bandwidth is comparable to the bandwidth of a one core Atom node.

The last two lines in Table 6.1 represent the SMXV measurements made on the most recent GeForce GTX 280 node by NVIDIA. Using the main memory that goes up to 4 GB in 2009, the double precision SMXV performance went up to 10 GF/s with a top main memory bandwidth of 15 Gw/s. The very high double precision R_a^{SMXV} value of 16 GF/s indicated in the second row is due to the usage of the texture cache of 1 GB. In these measurements it is supposed that the data already sits in main memory or in the texture cache, and can directly be attached from there by, for instance, an iterative matrix solver.

For the Nvidia GeForce GTX 280, the DGEMM measurement in Table 6.1 comes from the Internet. It indicates 74 GF/s; 95% of peak.

6.2.1 Race condition with OpenMP

The OpenMP directives shown in Fig. 6.13 did accelerate the execution by a factor of 3 when 4 cores are used. In both cases the outermost loop was the k counter. As a consequence, each thread produced worked on separate chunks of the resulting matrix c, and the results were right.

Table 6.3 OpenMP measurements on a single processor quadri-core Nehalem

N_{cores}	R_a GF/s	Result s
1	0.766	1'000'000'000
4	1.819	997'733'986
4	1.794	996'930'026

Let us consider the triple loop shown in Fig. 6.14. The outermost loop is now the i counter. What happens if we parallelize with OpenMP? Each thread acts on the entire resulting matrix c. The writing back of the results is in conflict with the other threads. As a consequence, race conditions occur. In Table 6.3 are shown results on the 4 cores Nehalem for the cases with one thread and with 4 threads. With one single thread, the result is correct. If the triple loop is executed demanding 4 threads distributed among the 4 cores, the results become wrong, and differently wrong when executing a second time. When applying OpenMP, we encourage the users to carefully check the coding and the results.

Let us also make here a remark on the race condition when using MPI. A message is sent with MPIsend to another core. There, an MPIreceive expects arrival of the message. The application programmer has to take care of the obtained message. He takes the responsibility to do the right move with the data. Thus, race conditions with MPI are improbable.

6.3 Application optimization with OpenMP: the 3D Helmholtz solver

Optimizing an application on a node or on a parallel machine is not very attractive. It does not contribute to the glory of the application user. He rather has to produce results that can be published. For this purpose, he uses resources that are available, if possible, for free. It is not his task to understand the program in detail, a must to optimize it on the more and more complex computer architectures. However, if the computations take a lot of time, the user has to wait results, and the productivity diminishes. Then, it can become interesting to accelerate the code to produce more results in a given time, or to run bigger cases with more physics in the model.

In some circumstances, the user is confronted with situations where he cannot always understand the results, or even results are wrong when he ports the code from one machine to another one. Sometimes the computing time increases a lot when changing the computational resources, then, the user believes that the new machine was a bad choice of the computing centre. By presenting a sort of "How to do" proceeding we try to show that optimization is a very useful one shot short time activity, often resulting in a substantial increase of productivity.

The vectorized Helmholtz solver runs at 85% efficiency on a NEC SX-5. The most time-consuming parts have been ported on SMP, NUMA, and cluster architectures. It is shown that an OpenMP version can deliver a similar performance when running on a 16 processor SGI Altix. A partial MPI parallelization uses the Γ model to predict application behaviors on parallel machines. This model is then applied to simulate the behavior of a hypothetical full MPI version on different distributed memory machines. It is found that only the Cray XT3 with its very fast internode communication network will be able to deliver the performance of a NEC SX-8 with the advantage that bigger cases could be handled. Since the efficiency on a Cray XT3 has been rather small, the MPI results are not presented. This application indeed needs a real shared main memory concept to run efficiently.

6.3.1 Fast Helmholtz solver for parallelepipedic geometries

In this section we focus on the fast diagonalization method for the solution of the elliptic Helmholtz problem with mixed Robin boundary conditions:

$$\Delta u - h_v u = f \qquad \text{on } \Omega \tag{6.1}$$

$$\alpha u + \beta \frac{\partial u}{\partial \mathbf{n}} = g \qquad \text{on } \partial\Omega, \tag{6.2}$$

where $\partial\Omega$ denotes the boundary of the open domain Ω. h_v is a positive real constant, α and β are real constants, the source terms f and g are given functions, and $\frac{\partial}{\partial \mathbf{n}}$ is the derivative along the normal direction to $\partial\Omega$. The computation is performed in a parallelepipedic, three-dimensional geometry, $\Omega =]-h, +h[^3$.

The space approximation is based on a mono-domain Chebyshev collocation spectral method (CCM). The spectral discretization proceeds by expanding the u field in a tensor product of high-order Lagrangian polynomials of order (N, M, L) for the (x, y, z) dependencies, respectively. In the present study, N, M, and L are equal, and denoted by N. The CCM consists of exactly enforcing the differential equations, and the boundary conditions, at the Gauss-Lobatto-Chebyshev points [59, 32].

Let us introduce several discrete operators: $\mathcal{L}$ is defined as the discrete Laplacian and $\mathcal{H}$ is the discrete Helmholtz operator, which includes the mixed Robin boundary conditions imposed on u. The discrete Helmholtz problem reads then as follows:

$$\mathcal{H}u = \mathcal{F}, \tag{6.3}$$

with

$$\mathcal{H} = \mathcal{L} - h_v \mathcal{I}, \tag{6.4}$$

and $\mathcal{I}$ is the unit matrix.

Let us define the one-dimensional matrices D_{xx}, D_{yy}, and D_{zz} as the discrete second order derivative matrices defined on the Gauss-Lobatto-Chebyshev points. In this way, the Helmholtz equation (6.3) can then be written as

$$(\mathcal{I}_z \otimes \mathcal{I}_y \otimes D_{xx} + \mathcal{I}_z \otimes D_{yy} \otimes \mathcal{I}_x + D_{zz} \otimes \mathcal{I}_y \otimes \mathcal{I}_x - h_v \mathcal{I}_z \otimes \mathcal{I}_y \otimes \mathcal{I}_x) u = \mathcal{F}. \quad (6.5)$$

with $A \otimes B \otimes C$ the tensor product of the one-dimensional matrices A, B, and C, and $\mathcal{I}$ the one-dimensional identity matrix in every space direction (x, y, z). It is clear that this operator is separable and invertible. Lynch et al. [100] showed that the inverse of such an invertible and separable operator $\mathcal{H}$ can be written as

$$\mathcal{H}^{-1} = P_z \otimes P_y \otimes P_x (\mathcal{I}_z \otimes \mathcal{I}_y \otimes \Lambda_x + \mathcal{I}_z \otimes \Lambda_y \otimes \mathcal{I}_x + \Lambda_z \otimes \mathcal{I}_y \otimes \mathcal{I}_x$$
$$- h_v \mathcal{I}_z \otimes \mathcal{I}_y \otimes \mathcal{I}_x)^{-1} P_z^{-1} \otimes P_y^{-1} \otimes P_x^{-1}. \quad (6.6)$$

P_x, P_y and P_z are the matrices involving the eigenvector decomposition of the one-dimensional operators D_{xx}, D_{yy} and D_{zz}, such that

$$P_x^{-1} D_{xx} P_x = \Lambda_x, \qquad P_y^{-1} D_{yy} P_y = \Lambda_y, \qquad P_z^{-1} D_{zz} P_z = \Lambda_z, \quad (6.7)$$

with Λ_x, Λ_y and Λ_z diagonal matrices with as entries the eigenvalues of the corresponding operators. The interest of this method, which is often called the fast diagonalization method (FDM), becomes clear when the cost of an evaluation of $\mathcal{H}^{-1} u$ is compared to the matrix-vector multiplication $\mathcal{H} u$, the basic ingredient of any iterative method. $\mathcal{H}^{-1} u$ requires $6N^4 + N^3$ multiplications, whereas $\mathcal{H} u$ takes $3N^4$ multiplications. So the inverse is computed at the price of two matrix-vector products. Many authors have used this fast diagonalization technique in the context of mono-element spectral methods, e.g. [71, 95, 97]. It is not surprising that this fast diagonalization method is often used in the context of spectral methods, where iterative methods tend to converge slowly due to the ill-conditioned matrices. Unfortunately, the condition that the discrete problem should be separable restricts the field of applications. For example, operators resulting from multi-domain discretizations can in general not be separated.

The numerical simulation of incompressible viscous Newtonian fluid flows is performed by solving the Navier-Stokes equations. In the framework of the spectral approximations, the state-of-the-art time algorithms handle the pressure and the viscous terms implicitly through an unsteady Stokes problem, with the non-linear terms treated explicitly in time as source. This holds even for the direct numerical simulation (DNS) of turbulent flows [96]. It turns out that per time step for a time marching scheme, three elliptic Helmholtz problems (for each component of the velocity) and one Poisson problem (for the pressure) need to be solved. In order to produce meaningful statistics, the number of time steps is typically of the order of 1,000,000. Solving accurately and efficiently the three-dimensional (3D) Helmholtz problem is therefore the corner stone of accurate and efficient numerical experiments.

6.3.2 NEC SX-5 reference benchmark

The Helmholtz fast diagonalization method has been executed for more than 10 years on the NEC vector machines that were installed at the Swiss National Computing Center (CSCS). The performance measurements on the NEC SX-5 are used as reference data for the comparison with the performances measured on the other machines. The data chosen for the benchmark execution is given by the polynomial degree N=169 in all three spatial directions and by the number of time steps nt=10 chosen to have a better evaluation of the behavior of the code.

After execution on one processor, the NEC machine delivered a profile:

```
       ******  Program Information  ******
Real Time (sec)          :          74.607827
User Time (sec)          :          72.193899
Sys  Time (sec)          :           0.417571
Vector Time (sec)        :          70.414532
Inst. Count              :        6486819207.
V. Inst. Count           :        3536376703.
V. Element Count         :      763644076883.
FLOP Count               :      490967420657.
MOPS                     :       10618.549861
MFLOPS                   :        6800.677417
VLEN                     :         215.939687
V. Op. Ratio (%)         :          99.615123
Memory Size (MB)         :         304.031250
MIPS                     :          89.852734
I-Cache (sec)            :           0.130809
O-Cache (sec)            :           0.210814
Bank (sec)               :           0.000315
```

One can deduce that 10 iteration steps take about 72 seconds for the 4.9×10^{11} floating point operations. Each diagonalization step corresponds to five DGEMM and four quadruple loops. Thus, one diagonalization step takes about 1.44 seconds on the NEC SX-5. Note that the memory size is about 300 MB, thus the program can be executed on any single processor machine. The performance on the SX-5 is 6.8 GF/s, corresponding to an impressive 85% of peak. We have to mention here that for a realistic problem the number of time steps of a full Navier-Stokes simulation is typically 1'000'000, leading to a total execution time of 3 months. On a newer NEC SX-8 machine such a case would run about two times faster. The question we ask: is there a single processor, a SMP or NUMA machine or an HPC cluster that is able to run such a case at least as fast as on a NEC SX-8?

6.3.3 Single processor benchmarks

The Helmholtz solver has been optimized on vector machines that demand long
vector operations. We had to adapt it to the RISC machines on which the right
compiler and compiler options have been chosen to get highest performance. For
instance, modifications on the loop orders have improved the execution times of the
four M4_LOOP subroutines.
    The profiling

```
Each sample counts as 0.01 seconds.
   %    cumulative    self
  time    seconds    seconds    name
 58.72    370.27     370.27     helm3col_
 12.48    448.99      78.72     M4_LOOP
 10.56    515.59      66.60     M4_LOOP
  3.54    537.94      22.35     M4_LOOP
  2.56    554.07      16.13     M4_LOOP
  1.55    563.83       9.76     _aligned_loop
  1.41    572.69       8.86     permzy_
  1.41    581.55       8.86     permyz_
  1.34    590.01       8.46     mkl_blas_p4n_dgemm_copyan
  1.23    597.78       7.77     MAIN__
  1.19    605.29       7.51     _one_steps_loop
```

shows that 58% of the time is consumed by the helm3col subroutine in which the
LAPACK subroutine DGEMM is called 5 times.
    We start the benchmarking with a set of low-cost single processor machines pre-
sented in Table 6.4. The first two lines show two results on a Pentium 4, the first
without and the second with hand-made optimization. All the other results are after
optimization. In the last column are listed the estimated total time in months. It is
expected that one result can be obtained every 1.5 months as on a NEC SX-8. We
see that for single processor machines the time to run one full simulation varies from
over half a year to one full year. Since efficiency is highest with a single node SMP

Table 6.4 Benchmarks on different CPUs. The NEC SX-8 data is estimated. R_∞ is the peak performance of the processor and M_∞ is the peak main memory bandwidth in Gw/s with 64 bit words.

Machine	f (GHz)	R_∞ (GF/s)	M_∞ (Gw/s)	t (s)	% of peak	T (months)
Pentium 4	2.8	5.6	0.8	6.15	28	12.7
Pentium 4	2.8	5.6	0.8	4.37	39	9.1
Opteron	2.4	4.8	0.8	4.82	46	10
Xeon	2.8	5.6	0.8	3.42	51	7.1
Itanium 2	1.3	5.2	0.8	3.26	58	6.8
NEC SX-5	0.5	8	8	1.44	85	3
NEC SX-8	1	16	8	0.72	85	1.5

machine, the CPU time costs are much smaller, but the waiting time is by far too long. As a consequence, single processor RISC machines cannot run the Helmholtz problem in the 1.5 months time frame.

6.3.4 Parallelization with OpenMP

The Helmholtz program has been parallelized using the OpenMP directives for SMP machines. This is an easy way to introduce the parallelism, but it is only possible if the memory is shared by the processors. In this version, DGEMM as well as the major loops have been parallelized. The results can be found in Table 6.5. We see that only the 4 processors Itanium 2 SMP machine runs slightly faster than a single processor SX-5, none of these machines reaches the performance of the NEC SX-8.

6.3.5 Parallelization with MPI

To port and parallelize the code on the three clusters,

- Pleiades 2: 100 Xeon 64bit single ($N_{CPU} = 1$) processor nodes with a full GbE switch
- Mizar: 160 dual ($N_{CPU} = 2$) processor Opteron nodes with a full Myrinet switch
- Cray XT3: 64 single ($N_{CPU} = 1$) processor Opteron nodes with a Seastar 3D interconnect

we perform the following steps:

- Compile the code coming from NEC and improve efficiency
- Profile the code
- Optimize the sequential parts
- Parallelize DGEMM that takes 58% of the time
- Compare with the Γ model
- Apply the Γ model to simulate scalability of a fully parallel code.

Table 6.5 Benchmarks on different SMP machines using OpenMP

SMP	f (GHz)	#CPU	R_∞ (GF/s)	t (s)	% of peak	T (months)
Opteron	2.4	2	9.6	3.54	25	7.4
Itanium 2	1.3	2	10.4	2.13	38	4.4
Itanium 2	1.6	2	12.8	1.86	41	3.9
Itanium 2	1.3	4	20.8	1.38	31	2.9

6.3.5.1 Parallelize DGEMM with MPI

The matrix*matrix operation C=A*B takes most of the CPU time. The matrix A has been distributed on all computational nodes, whereas B has been cut in slices, each slice has been sent to the corresponding node. Each partial matrix*matrix multiplication is performed calling DGEMM. The partial results of C are collected and combined.

The measured speedups as a function of the number of nodes are shown in the upper Fig. 6.15, and compared with estimated ones at the bottom. The Pleiades 2 cluster with a GbE switch shows speed-down. When using more than one processor, the Mizar machine including a Myrinet switch shows speedup up to the two processors in a SMP node, then there is no additional speedup. The only cluster that shows a considerable speedup is the Cray XT3 having a very powerful internode communication system.

These speedup curves have also been simulated using the Γ model. In this model, the application is parameterized by the quantities V_a for DGEMM ($V_a^d = 10$) and for the loops ($V_a^l = 4$). V_a measures the number of operations performed for every word written to or read from the main memory.

The Γ model is used explicitly to forecast the running time. First one has to calculate r_a^d for the parallel DGEMM part

$$r_a^d = min(R_\infty, M_\infty * V_a^d) = R_\infty$$

and r_a^l for the non parallelised operations:

$$r_a^l = min(R_\infty, M_\infty * V_a^l) = M_\infty * V_a^l$$

then one can calculate the computation time for one DGEMM parallelised routine which depends on the number of processors and the size of the matrices:

$$t_c(P, N) = \frac{(2 * N^4)}{(P * r_a^d)}$$

the communication time:

$$t_b(P, N) = \frac{((2 * N^3 + N^2) * P * max(0, P - p)* < d >)}{(C_\infty * P)}.$$

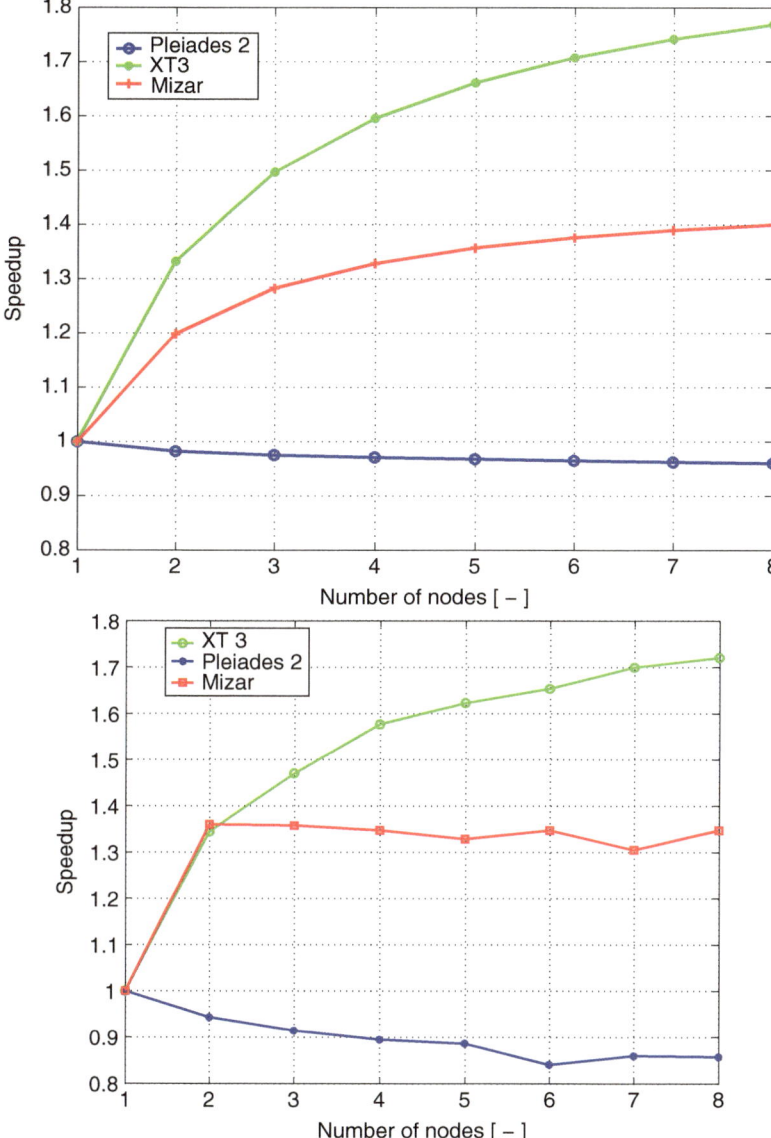

Fig. 6.15 Top: Estimation of speedup. **Bottom:** Measured speedup

The total time t_{total} corresponds to the 5 DGEMM calls plus the non parallelized operations time (M):

$$t_{total}(P, N) = 5 * (t_c(P, N) + t_b(P, N)) + \frac{M}{(P * r_a^l)}$$

6.3.5.2 Simulation of a fully parallelized code

The Γ model can now be applied to simulate a fully parallelized Helmholtz solver without doing the cumbersome work of data distribution. The results are presented in Fig. 6.16. The upper curves are for the Cray XT3, the middle ones for Mizar having two processors nodes (but simulated as single processor nodes with half network bandwidth), and the lower curves for Pleiades 2. One sees that only the Cray XT3 machine gives a reasonable speedup A. This speedup levels off at $A = 21$. The single processor CPU time of one Cray node has been measured to take 5 seconds for one step, just 7 times longer that the estimated 0.72 seconds of a NEC SX-8. With this measurement, one can state that 10 processors Cray XT3 (upper most curve) will deliver the performance of one processor NEC SX-8. However, the Helmholtz solver has first to be parallelized entirely, implying a distribution of the data among the processors and programming the data exchange using the MPI library. This effort has to be compared with the lower execution costs and a decision on programming investment with respect to computing investment must be made.

The SGI Altix NUMA cluster running at 1.6 GHz has also been simulated (second curve from above) and added in Figure 6.16. It is predicted to deliver a 4 fold speedup with 16 processors. The corresponding performance is then comparable with the one of a NEC SX-8

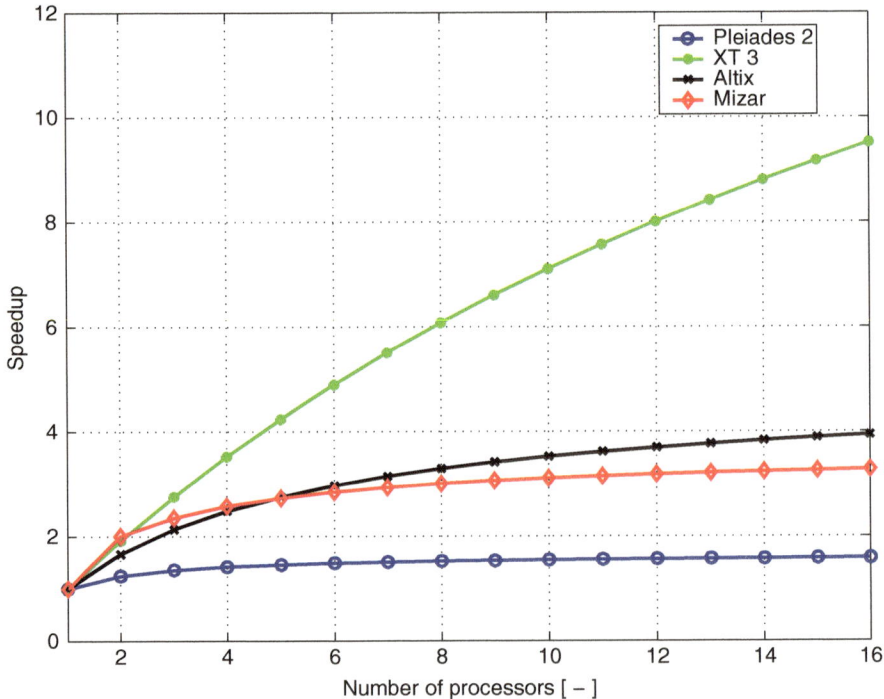

Fig. 6.16 Estimation of the speedup for a fully parallelized code

6.3.6 Conclusion

The Helmholtz solver is a highly communication intensive program. To get high performance it is needed to run the code on a high performance vector processor machine like NEC SX-8 or on a machine having a very high internode communication bandwidth. Specifically, to reach the performance of the NEC SX-8 it is needed to run on 16 processors of the SGI Altix NUMA machine, or on 10 processors of the Cray XT3. For the NUMA machine the code has to be turned into an OpenMP code. To run on the Cray XT3, it is necessary to produce an MPI based code.

6.4 Application optimization with OpenMP: TERPSICHORE

In the previous chapter, the optimization of TERPSICHORE on one core has been presented. We have seen that the compiler can perform an almost perfect code optimization, with hand optimization giving an only slightly faster program. However, if one runs a compiler optimized program on a multi-core node, and adds OpenMP directives, these directives can interfere with the compiler optimization process, and the parallel efficiency is small, sometimes even negative. Let us present a few measurements.

In Table 6.6 are presented measurements made on an Intel Xeon, also called Woodcrest 5160, having two processors, two cores each. The total peak performance is 21.3 GF/s, and 42.6 GF/s for the SSE acceleration hardware. The node is in the first column, and the second column indicates the compiler options chosen, -O1 means no optimization whatsoever, -O3 is a high optimization option that can be made more aggressive by choosing in addition -xT or even -xT -parallel.

The third column shows the CPU times on a single core (first 3 results) and with option -parallel (fourth result) for a test case that took 30 seconds on the NEC SX-5 vector machine. With the choice of the automatic options without hand optimization TERPSICHORE could be accelerated by a factor close to 2 (from 133 s to 65 s), reaching almost half of the NEC SX-5 speed. When optimizing by hand and without switching on OpenMP, it is possible to get a somewhat better result, the 65 seconds best can be pushed down to 51 seconds.

When switching on OpenMP on top of the hand optimized code, it is interesting to see that aggressive compiler options have almost no effect on the performance of

Table 6.6 TERPSICHORE: CPU time measurements on a Woodcrest 5160 using ifort compiler, and on a NEC SX-5

Computer	Compiler options	*no hand optimization* s	*by hand* s	*with OpenMP* s
Woodcrest	-O1	133	85	31
Woodcrest	-O3	124	84	31
Woodcrest	-O3 -xT	96	84	28
Woodcrest	-O3 -xT -parallel	65	51	27
NEC SX-5		30		

Table 6.7 TERPSICHORE: CPU time measurements on a Woodcrest 5160 node in the Pleiades cluster running at 2.67 GHz

Compiler options	Cluster node	Speedup
-O3 -xT	84	1
-O3 -xT -parallel	51	1.70
-O3 -xT -openmp 1	80	1.05
-O3 -xT -openmp 2	45	1.88
-O3 -xT -openmp 3	36	2.33
-O3 -xT -openmp 4	28	3.00

the parallel code: 31 seconds for -O1, 27 seconds for the most aggressive option. One can realize that the best CPU time is slightly smaller than the one measured on the NEC SX-5. However, the efficiency on the NEC SX-5 was 51%, whereas on the Woodcrest, TERPSICHORE reaches a high 10.6% RISC machine efficiency. This is due to the important DGEMM part in the eigenvalue solver.

In Table 6.7 are compared OpenMP measurements with the hand optimized version and the best single core compiler options -O3 -xT for which 84 seconds have been measured. Then, three of the four cores were idle. With the option -parallel, the compiler automatically parallelizes, reducing the execution time by 40%. The measurements with OpenMP show improvements by up to a factor of 3 when using all the 4 cores.

Chapter 7
Cluster optimization

"Don't blame the messenger because the message is unpleasant."

Kenneth Starr, American lawyer

Abstract After having optimized a code at a core-level (Chapter 5) and at a node-level (Chapter 6), the user may want to parallelize his application at a cluster level, the purpose of this chapter. The authors are aware that parallelizing a sequential application can lead to a high number of problems, and even to sometimes wrong results if a race condition is introduced. Parallelization can be mandatory if a single nodes does not offer enough main memory or if its performance is insufficient.

7.1 Introduction on parallelization

We start with a presentation of internode communication networks, followed by distributed memory cluster architectures. The different type of applications are discussed, the differences are related to the different levels of network communication needs. A specific domain decomposition technique is then presented in more detail. In this technique, the geometrical domain is cut in subdomains, each subdomain is taken care of by one virtual processor. Connectivity conditions interconnect the subdomains. Examples are shown coming from plasma physics, engineering, and computational fluid dynamics (CFD).

7.2 Internode communication networks

7.2.1 Network architectures

Let us look into the most popular network architectures that can be used to interconnect NUMA and distributed memory architectures.

7.2.1.1 The Bus

The most common communication system is the Bus with one link, $\ell = 1$, and a bandwidth of $C = C_\infty$, Fig. 7.1. Altogether N_{node} computational nodes are

R. Gruber, V. Keller, *HPC@Green IT,*
DOI 10.1007/978-3-642-01789-6_7, © Springer-Verlag Berlin Heidelberg 2010

Fig. 7.1 The Bus
communication architecture

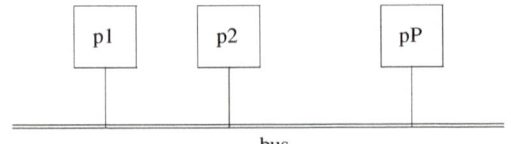

connected to one cable, the Bus. If one message is sent to all the nodes, a Bus communication is efficient. If different nodes send different messages to different destinations, the overall bandwidth of the Bus has to be shared among all the nodes, and the effective bandwidth per node can be as small as C_∞/N_{node}. On the other hand, the overall cost of a Bus is independent of the number of nodes, if one does not count the costs of the node connections. We shall see that there are a few parallel scientific applications for which a Bus offers sufficient bandwidth.

7.2.1.2 The Full connectivity

The other extreme of a Bus is a Full connectivity between all the nodes. As shown in Fig. 7.2 for the case of $N_{node} = 8$, each node is directly linked to every other nodes. The overall communication bandwidth per node is $C_\infty(N_{node} - 1)$. The bisectional bandwidth (total bandwidth between two halfs of nodes) is given by $H = C_\infty N_{node}^2/4$, $H = 16C_\infty$ for the configuration of Fig. 7.2. Thus, all the nodes can send at the same time a message to all the other nodes, and the communication time is the same as to send one message to a neighbor. A full connectivity network is the ideal situation, but the costs grow quadratically with the number of nodes. This is the reason why fully connected networks are very rare.

7.2.1.3 The Switch and the Fat Tree

The bisectional bandwidth H of a crossbar Switch or of a Fat Tree (Fig. 7.3) is equal to $H = C_\infty N_{node}/2$. This corresponds to the geometrical mean between a Bus with $H = C_\infty$ and a Full connectivity network (Fig. 7.2) with $H = C_\infty N_{node}^2/4$. In a

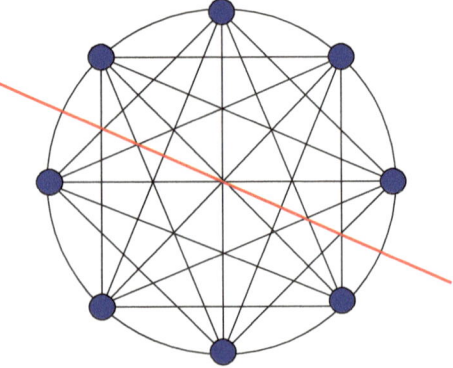

Fig. 7.2 The full connectivity network. The bisectional bandwidth is defined as the bandwidth through the number of links (here 16) cut by a mid plane

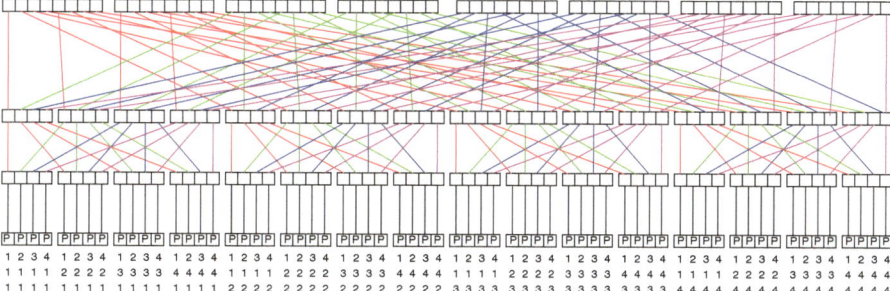

Fig. 7.3 A 64 nodes Fat Tree made with 8 × 8 crossbar Switches. There are three level of switches. The first level connects half of the links to the nodes, and half of the links connects to the second level interconnecting Switches. The third level interconnects the Switches of the second level. Four paths are possible per level, altogether 64 different paths are possible between two nodes

Switch or in a Fat Tree, a node has exactly one link to the network. Since the links are in general bidirectional, a node can send one message and receive one message at a time.

The Fat Tree [94] consists of a number of standard crossbar Switches with ℓ links each. In the Fat Tree represented in Fig. 7.3, half of the lowest level links of the 16 8 × 8 Switches connect to the $N_{node} = 64$ computational nodes. The other 64 links connect to the second level 16 8 × 8 crossbar Switches, and finally, those are interconnected by eight 8 × 8 third level Switches. The number of levels grows with the logarithm of the number of nodes. This Fat Tree network that connects 64 nodes consists of 40 8 × 8 Switches and 128, in general bidirectional links. The Fat Tree in Fig. 7.3, has 4 communication paths per level, altogether 64 different paths between two nodes. Bottlenecks can happen, but are rare. Examples of Fat Trees are those built by Quadrics, Myrinet and Infiniband.

7.2.1.4 The Grid and Torus network architectures

Let us remark that the designation "Grid" is used in many different ways. It can designate a geometrical grid on which a partial differential operator is discretised (also called a mesh), a number of machines or clusters that are interconnected in a transparent manner for the user (often written as GRID), as discussed in chapter 8, or, as presented here, it also designates a communication network (see Fig. 7.4). Two-dimensional (2D) or three-dimensional (3D) Grid networks are well adapted to solve iteratively partial differential equations in a structured numerical mesh. If the geometrical domain can be decomposed in structured subdomains with the same numbering as the underlying Grid network, communication has only to be done between nearest neighbors. If the subdomain numbering does not follow the network, the communication paths through the network can become long. In the network shown in Fig. 7.4, the distance, defined by the longest path between the nodes, is equal to $D = 8$. The average distance is $< D > = 3.2$. This corresponds

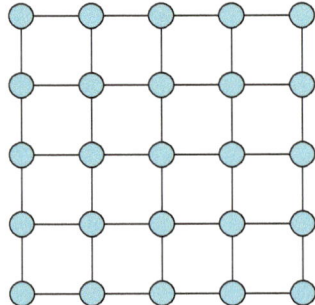

Fig. 7.4 A two-dimensional Grid network. Each node has four nearest neighbors to which they are directly connected. The number of steps through the networks grows with the number of nodes in the network, and the effective per processor bandwidth diminishes. Nodes sitting on extreme edges of the Mesh have longest distances. $D=8$, $< D > =3.2$

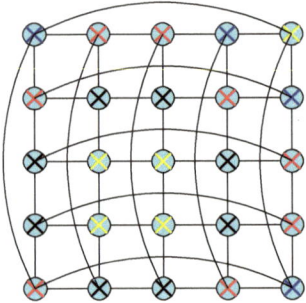

Fig. 7.5 A two-dimensional Torus network. As in the Grid, the nearest neighbours are interconnected. The difference is the periodicity of the Torus architecture. It is thus possible to send a message in two directions and the distances can be reduced by up to a factor of 2. $D=4$, $< D > =2.4$

to the weighted average of all the average distances of each node to all the other nodes.

The distance and the average distance can be reduced by a Torus network (see Fig. 7.5). Due to the periodicity of the network, the distance has dropped from 8 to $D=4$. Each node has the same average distance in the network. With the colouring of the nodes in Fig. 7.5 it can be shown how such an average distance can be computed. Let us start from the upper right point. The 4 points indicated with a blue cross have a distance of 1, the red crossed nodes have a distance of 2, the black crossed nodes a distance of 3 and, finally, the yellow crossed points have a distance of 4. The average distance then is

$$< d >= (4 \times 1 + 8 \times 2 + 8 \times 3 + 4 \times 4)/25 = 3.2,$$

and the average bandwidth per link becomes $b=C_\infty/< d >$.

Torus networks are often used by computer manufacturers since it can easily be built. Cray has started with their high bandwidth 3D Torus architecture machines T3D and T3E. They could go up to $4096 = 16 \times 16 \times 16$ nodes with a maximum distance of $D = 21$. Now, Cray Inc. builds the Red Storm machine (Cray XT3) with 6 links per node and a maximal bidirectional bandwidth of 7.6 GB/s per link. The trend is going to nodes that include on-chip communication cards for Grid and Torus networks. We have to mention here that a Torus network can become a Grid if one only uses a partition of directly interconnected nodes.

7.2.1.5 The Hypercube, the K-Ring, and the Circulant Graph

Different Hypercubes are represented in Fig. 7.6. The number of nodes $N_{node} = 2^D$ grows exponentially with the distance D. The average distance in a hypercube is

$$< d > = \frac{D 2^{D-1}}{2^D - 1}. \tag{7.1}$$

This architecture has been used in the SGI Origin machines in which so-called express links (see Fig. 7.10) have been added to reduce the distance by 1. Such an architecture is identically to the K-Ring architecture [91] used in the Swiss-T1

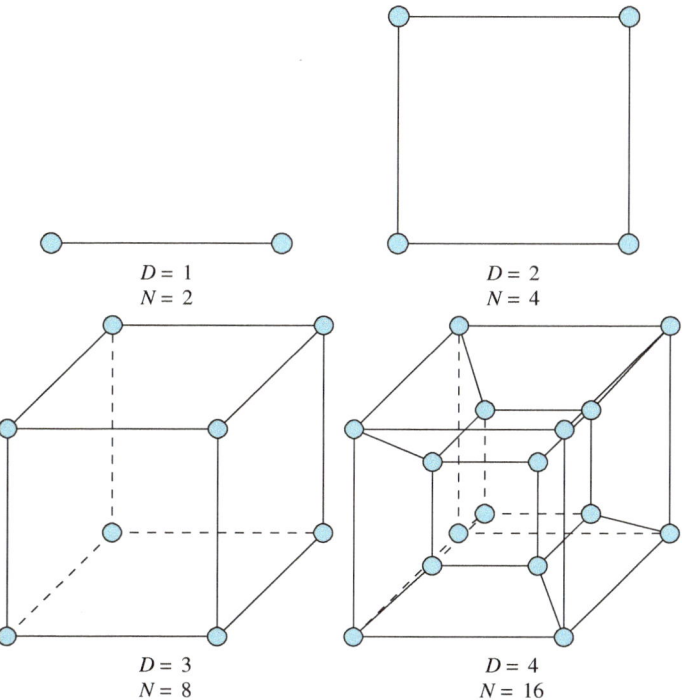

Fig. 7.6 The Hypercube architecture for D=1, 2, 3, and 4. The number of nodes is $N_{node} = 2^D$

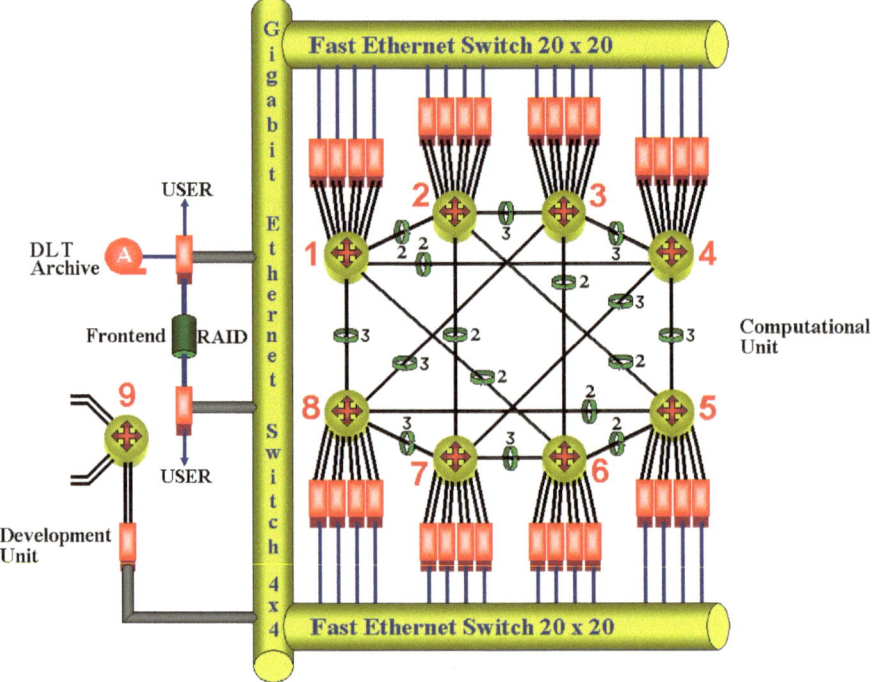

Fig. 7.7 The architecture of the Swiss-T1 with its eight 12×12 crossbars and the Fast Ethernet switch connections. The numbers on the links of the Circulant Graph network architecture denote the number of data transfers for an all-to-all operation between the crossbars

cluster, Fig. 7.7. Specifically, the (1 to 8) numbering in the Hypercube corresponds to the (1, 2, 5, 6, 3, 8, 7, 4) numbering in the K Ring.

A K-Ring is characterized by the number of nodes, N_{node}, by the number of rings, K, by the jumps of each ring, A_k, k=1, .., K, and by the full connectivity, i.e. the greatest common denominator gcd(A_k, k=1, .., K, N_{node})=1. In a ring, one numbers all the nodes from 1 to N_{node}, connects node number A_1 with node A_{k+1}, then A_{k+1} is connected with A_{2k+1}, then A_{2k+1} is connected with A_{3k+1}, and so on. With the restriction of gcd(A_k, k=1, .., K, N_{node})=1, a Ring visits all the nodes before it comes back to node number A_1. In Fig. 7.10 at the right is presented a K-Ring with (K, N_{node}, D, A_1, A_K)=(2, 8, 2, 1, 3), and in Fig. 7.9 one with (K, N_{node}, D, A_1, A_2, A_K) = (3, 21, 2, 1, 5, 8). In both cases, D=2, and the average distances are $< D > =1.25$ and $< D > =1.62$, respectively. The number of links per nodes is $2K$.

A more general Circulant Graph network [19] is characterized by the number of nodes, N_{node} in a cluster, numbered from 1 to P, by the number K of graphs, and by the jumps between two nodes A_k, k=1, .., K. A Circulant Graph is connex if gcd(A_k, k=1, .., K, N_{node})=1, (corresponding to a K-Ring) or not connex if gcd(A_k, k=1,..,K, N_{node}) $\neq$ 1. An example of such a non-connex graph is

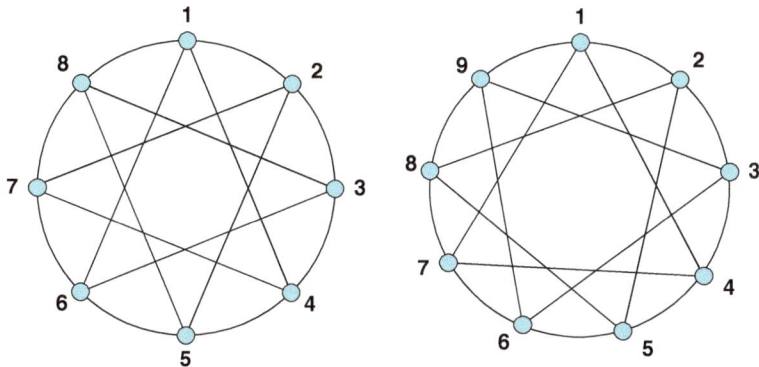

Fig. 7.8 From the K Ring with $N_{node}=8$, $K=2$, $A_2=3$, $D=2$, $<D> =1.25$ to the Circulant Graph with $P=9$, $K=2$, $A_2=3$, $D=2$, $<D> =1.33$

$(K, N_{node}, D, A_1, A_K)=(2,9,2,1,3)$ and $< D > =1.33$ presented in Fig. 7.8 at the right. There can be different graphs with the same optimal distance, for instance for $K=3$, $D=2$, the maximal number of nodes is 21 with 6 different optimal 3-Rings, $(3,21,2,1,2,8)$, $(3,21,2,1,3,8)$, $(3,21,2,1,4,6)$, $(3,21,2,1,4,10)$, $(3,21,2,1,5,8)$, $(3,21,2,1,5,9)$, all with $< D > =1.62$. There are other magic number of nodes together with the K value for which a K-Ring has an optimal value of D. These are:

$(K, N_{node}, D, A_1,..,A_K)$
$(2,13,2,1,5)$
$(2,25,3,1,7)$
$(2,41,4,1,9)$
$(3,55,3,1,5,21)$
$(3,55,3,1,10,16)$
$(3,55,3,1,20,24)$
$(3,105,4,1,5,41)$
$(3,105,4,1,7,29)$
$(3,105,4,1,16,22)$
$(3,105,4,1,38,46)$
$(3,105,4,1,43,47)$
$(4,35,2,1,6,7,10)$
$(4,35,2,1,7,11,16)$
$(4,104,3,1,16,20,27)$

The Fig. 7.9 shows the case $(3,21,2,1,5,8)$.

An advantage of a Circulant Graph to the Hypercube architecture with or without express links is its great flexibility since it can be built with any number of nodes. The K ring $(K, _{node}, D, A_1,A_K)=(2,8,2,1,3)$ shown in Fig. 7.10 has an identical connectivity as the Hypercube $D=3$ with express links bringing D down to $D=2$.

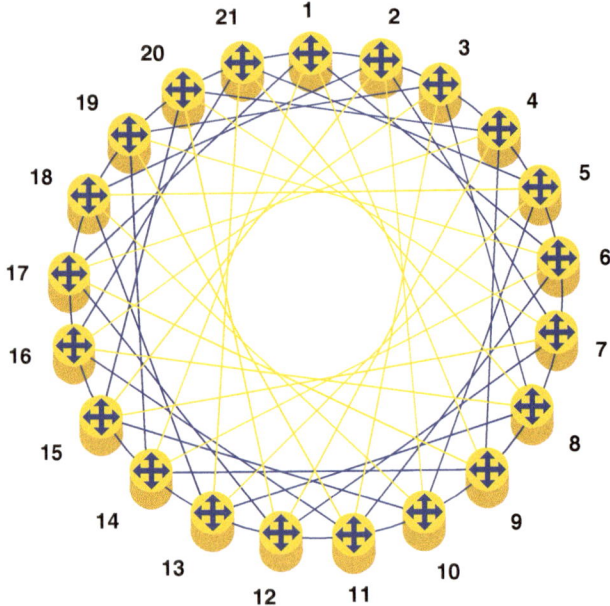

Fig. 7.9 An optimal K-Ring configuration. $K=3$, $P=21$, $D=2$, $<D>=1.62$, $A_2=5$, $A_3=8$

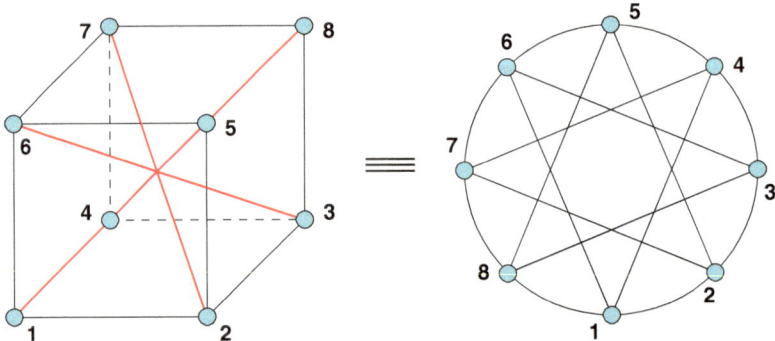

Fig. 7.10 The Hypercube architecture with diagonal express links. This architecture corresponds to the K Ring with $N_{node}=8$, $K=2$, $A_2=3$, $D=2$, $<D>=1.25$ chosen for the Swiss-T1, Fig. 7.7

The next Hypercube can be constructed for $D=4$, the number of nodes goes from $N_{node}=8$ to $N_{node}=16$. In the case of a Circulant Graph, the number of nodes can be augmented one by one. Thus, it is easy to go from the K ring in Fig. 7.10 to the Circulant Graph $(K, N_{node}, D, A_1, A_K)=(2,9,2,1,3)$, Fig. 7.8 in which the number of nodes has just increased from $N_{node}=8$ (left) to $N_{node}=9$ (right).

7.2.2 Comparison between network architectures

In Fig. 7.11 is represented the variation of the maximal distance D as a function of the number of nodes N_{node} for different network architectures. The upper most blue curve shows D for a 2D Torus, the green curve just below is for the 3D Torus, the 2 Ring is represented by the red oscillating curve, the 3 Ring by the blue oscillating curve, and for the Fat Tree the points for N_{node}=64, and 256 are indicated as circles and the black line connects to the circle at N_{node}=1024 that lies outside the figure. We can learn that a K Ring has a smaller distance than a KD Torus. The distance of a 3 Ring can be compared with the one of a Fat Tree. This means that point-to-point messages take about the same time in both networks. However, the distance is not the only characteristic of a network, the number of possible paths from one node to another one is bigger in a Fat Tree than in a 3 Ring. As a consequence, for a network with a large number of nodes, multicast messages will pass faster in a Fat Tree than in a 3 Ring.

7.2.2.1 Master/Slave network

In the Master/Slave network architecture the Master node distributes tasks to the different Slave nodes, Fig. 7.12. The Slaves perform the computations and send the results back to the Master node. The Slave nodes do not interoperate. This concept demands a Master node that can deliver a high bandwidth to the Slaves. Typically, a Master node is a server with enough network bandwidth to the Slaves. It also takes care of the I/O to and from disks.

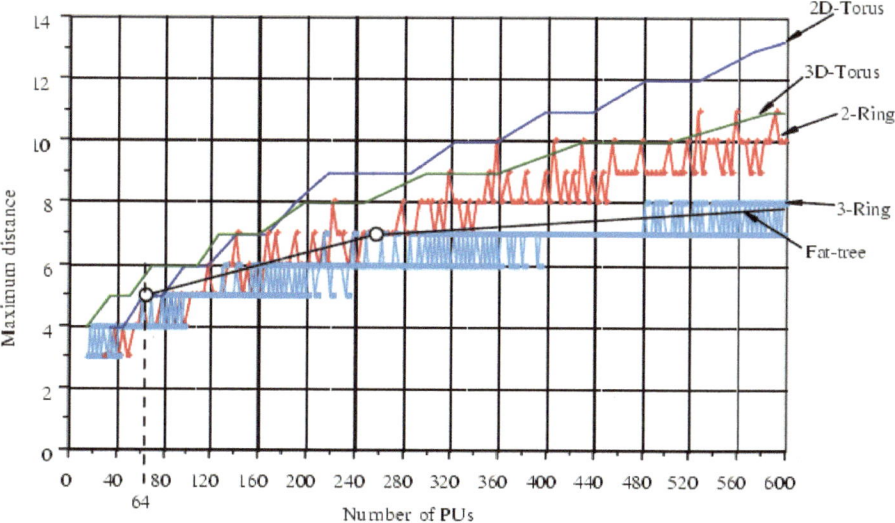

Fig. 7.11 Comparison between network architectures

Fig. 7.12 A Master/Slave concept. The Slaves are only connected to the Master

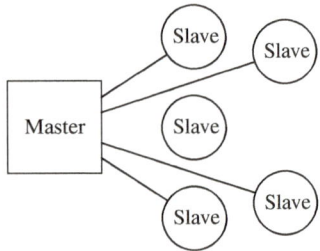

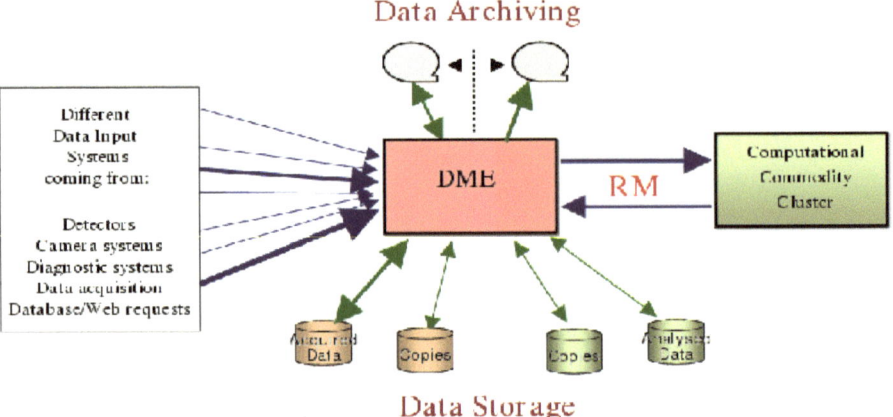

Fig. 7.13 A Master/Slave concept as it appears in practical situations. The Data Management Environment (DME) includes the so-called Frontend that takes care of data storage on disks and archive system of measured data. The data is interpreted on the computational unit consisting of a certain number of slaves

An example of a master-slave concept is shown in Fig. 7.13. Experimental data from detectors, camera systems, diagnostic and data acquisition systems, or requests from data bases, or to a Web server are sent by the master to a free slave node for computation. The slave sends the result back to the master. The master must be a secure environment, whereas the slaves can fail. When a slave fails, the computation is repeated, i.e. the master resubmits the case to an other slave. Such situations can be found in life sciences where data comes from mass spectrometers (protein sequencing). Each spectrometer produces independent data about a protein. The master, or the Data Management Environment (DME) first archives the data on disk and on cartridge, then submits it to a certain number of nodes of the Computational Commodity Cluster (CCC) for computation. When one of the nodes dies, the DME cleans up the other nodes implied in the computation of the same protein and resubmits the case to another partition of the CCC. This scheduling has successfully been implemented to a 1420 processor cluster for a bioinformatics company.

7.3 Distributed memory parallel computer architectures

7.3.1 Integrated parallel computer architectures

These parallel machines are characterized by an integration of the computational nodes with the underlying network architecture. They are often referred as MPP (Massively Parallel Processors). The computer vendor delivers a machine consisting of specific nodes, and an integrated internode communication network with a huge bandwidth to satisfy the needs of the applications. The operating systems are often Unix based as Linux, and the compilers and communication libraries are adapted to the underlying hardware. These HPC machines are installed in computing centers in which all types of user applications have to be supported. Let us present a few of those MPP machines (see Table 7.1).

7.3.1.1 Earth Simulator (ES, 2002, NEC SX-6)

This machine shocked the American IT industry in 2002. From this date of installation until 2004, the ES machine was number 1 in the TOP500 list. This list measures the Linpack performance (dominated by full matrix times full matrix DGEMM multiplications having $V_a > 100$). If one would measure the performance with the sparse matrix times vector multiplication SMXV with $V_a=1$, see chapter 3, this first position would have lasted at least until the June 2008 list.

The ES consisted of 640 computational nodes, each node was a NEC SX-6 with 8 vector processors running at 500 MHz. Each processor had 16 arithmetic pipelines, giving altogether 81'920 arithmetic functional units. The peak performance was over 40 TF/s, there were 10 TB of main memory, 700 TB of disk space, 16 GB of bisectional inter-node communication bandwidth, covered $10'000\,m^2$ of space, is air cooled, and a newer version based on NEC SX-6 nodes consumes 3.2 MW. This machine was used for applications related to earth simulation such as simulations of earthquakes, tsunamis, tornados, for climate modeling, weather forecast, plasma physics, or astrophysics. In 2009, the ES was still in production.

Table 7.1 TF/s rates of some vendor specific MPP machines

Machine	ES	Blue Gene	XT3	Altix	p-series	X1E
Location	Tokyo	LLNL	SNL	NASA	LLNL	Korea
Vendor	NEC	IBM	Cray	SGI	IBM	Cray
Procs	SX-6	Power 4	Opteron	Itanium	Power 5	
Nodes	640	65'536	10'880	20	1280	1
P/node	8	2	1	512	8	1020
Memory	10	32	22	20	40	4
DGEMM	36	281	36	52	63	15
SMXV	29	35	5	7	8	12
OS	UX	AIX	UNICOS	IRIX	AIX	UNICOS

7.3.1.2 IBM Blue Gene (2004, LLNL)

The IBM Blue Gene/L installed at Lawrence Livermore National Laboratory was number one in the Top500 from 2004 to 2007. The machine consisted of 32'768 (2004) and 106'496 (2008) dual-core 700 MHz (2.8 Gflops/s) Power 4 processor nodes, Table 7.1, 512 per rack (see Fig. 7.14). One core per node can be used for computation, and one for communication. If communication is light, both cores can compute. The main memory per node is 512 MB. The whole memory is attached to the single computational core (if the other is only performing communication), or 256 MB are attached to each of the two cores. If more main memory is requested in a core, the job fails. The memory bandwidth of the 350 MHz bus is 2.8 GB/s for each of the two cores. Different networks are available on a Blue Gene: 3D torus network for point-to-point MPI operations, a fat tree network for multicast MPI operations, a bit-wise master/slave network for barriers, and a network for administration. The power consumption per core is 11 W (see Fig. 2.2). Newer machines, i.e. the Blue Gene/P, run at 850 MHz and consume 7.7 W per core.

7.3.1.3 Cray XT3, Redstorm (2005, SNL)

This machine has been designed in a co-operation between Sandia National Laboratory and Cray Inc. The design goal was to build an internode communication network system that is simple (3D Torus network) and scalable up to one million of processors. The first machine installed at SNL, Table 7.1, consisted of 10'368 single processor/dual core AMD Opteron compute nodes, running

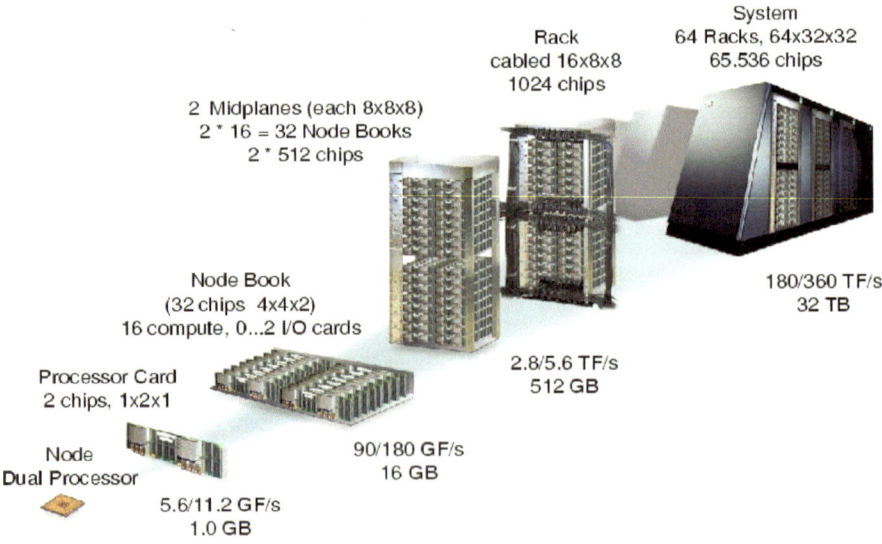

Fig. 7.14 The hierarchical architecture of the IBM Blue Gene machine

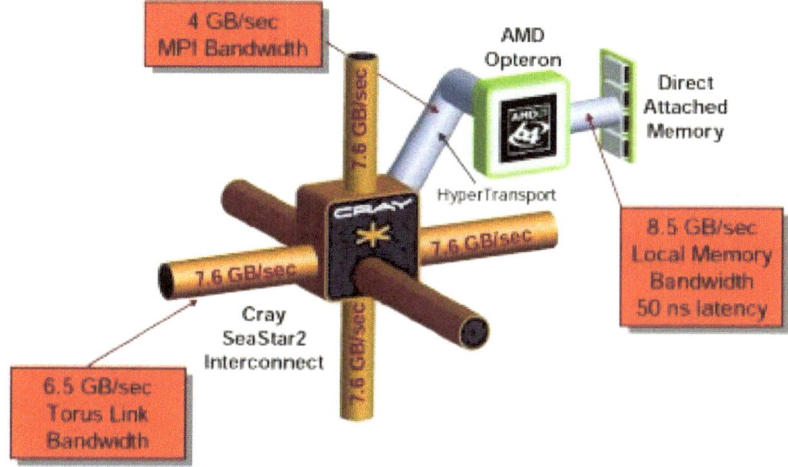

Fig. 7.15 The network component that interconnects adjacent nodes to a 3D Torus in the Cray XT3 Scalable Architecture.

at 2.2 GHz, connected to a 3D mesh/torus inter-node communication network. Each link has a bidirectional bandwidth of 7.6 GB/s (see Fig. 7.15), comparable with the local main memory bandwidth of a node. The system is managed and used as a single system, an easy to use common programming environment (shmem), a high performance distributed file system LUSTRE, a special low latency, and a very high bandwidth message passing library are available. The user can reserve a fully connected local partitions. The XT3 is liquid cooled and takes only $300\,m^2$.

A newer petaflop version has been installed in Oak Ridge Laboratory in 2008 with 37'538 quad core AMD Opterons running at 2.3 GHz with an energy consumption per core of 46 W leading to a total of close to 7 MW. In the November 2008 TOP500 list, this machine was top in energy consumption.

7.3.1.4 SGI Altix (2004, NASA)

SGI has decided to stop their Origin series that were appreciated because of the powerful user-friendly environment with best rated compilers. They switched to Itanium processors with Intel compilers. We have seen that the most recent *ifort* and *icc* compilers for Itanium give very high performance executables. The NUMA architecture and the Numalink inter-node communication up to 512 processors have been ported to the new Altix machines. Machines with more than 512 processors must be interlinked with other communication systems such as a Voltaire Infiniband switch as the one on the NASA machine (see Table 7.1). OpenMP applications can run on a node with up to 512 processors. If more processors are needed, MPI or a mixture between OpenMP and MPI have to be used. In 2008, the new Altix machine at NASA had altogether 51'200 cores based on 3 GHz Intel Harpertown nodes with a per core energy consumption of 40 W.

7.3.1.5 IBM p-series

The IBM p-series machines are clusters of Power 6 processors based nodes with up to 32 processors. The two machines installed at ECMWF in 2008, each consists of 262 nodes with 8 processors each (see Table 7.1) running at 4.7 GHz. The per processor energy consumption is 160 W.

7.3.1.6 SUN Blades, 2008

The half petaflop Ranger machine at the Texas Advanced Computing Center (TACC) had 15'744 blades of four quad core Opterons running at 2.6 GHz. The energy consumption was 30 W per core.

7.3.1.7 Roadrunner, 2008

The first big machine, the Roadrunner at Los Alamos National Laboratory (number 1 in the November 2008 TOP500 list), includes 20'000 Dual processor Cell; each is attached to two AMD dual core Opterons. This leads to 1.456 PF/s peak performance, and to over one PF/s performance for the Linpack benchmark. In fact, this was the first PF machine. The energy consumption per Cell processor is 200 W, its peak performance is just over 100 GF/s, leading to 2 W per GF/s. This is slightly better than a Blue Gene/P (7.7 W for 3.2 GF/s).

7.3.2 Commodity cluster architectures

A commodity cluster consists of series of mass produced hardware modules and software that can be downloaded from the Web. An example is the Pleiades2 cluster (pleiades2.epfl.ch). This parallel machine consists of a secure frontend environment including 2 dual processor/dual power supply servers with a Raid disk system and a tape robot, and 220 dual processor Intel Woodcrest nodes, each processor has 2 cores. Each node has 2 times 4 GB of main memory each having a main memory bandwidth of 10.2 GB/s, a graphic card, a local disk of 80 GB, and a Floppy disk reader, thus, just the parts needed for a computational node. A full GbE switch connects all the modules of this cluster.

Linux is used as operating system, the resource management system OpenPBS and the Maui scheduler take care of job submission, NFS of the file handling, Systemimager is used to install software on all the nodes in parallel. The Fortran, C, and C++ compilers are from Intel and GNU, MKL, FFTW, Mumps, Petsc, Arpack, Scalapack are the mathematical libraries, MPI and PVM the communication libraries. A special PVFS (Parallel Virtual File System) and an eight node I/O system have been installed to run big I/O operations over the switching network and not through the frontend using NFS.

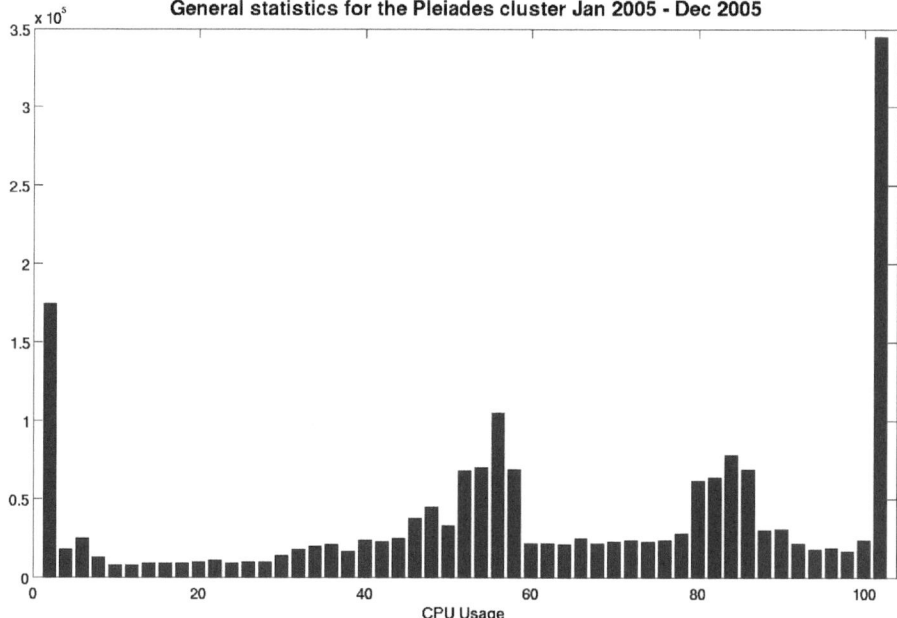

Fig. 7.16 CPU usage (in %) statistics over 12 months on the Pleiades 1 cluster with 132 processors as a function of the number of time units collected by Ganglia. The average usage is 64%

The machine runs 24 hours a day, 7 days per week. The CPU time used is 68% of all available cycles, Fig. 7.16. The remaining 32% are due to cores that are down, due to I/O operations during which the processor is idle, due to electricity or water cut downs, and due to blocked jobs. The dominant loss in cycle periods comes from reserved but not used cores. In fact, one can detect parallel applications in which only one core out of four is active during long periods of time.

An older version, Pleiades1, had 132 Pentium 4 processors, interconnected by a Fast Ethernet switch. We mention this machine because we learned that the bandwidth of this slow network was not adequate for a number of applications. In Fig. 7.17 are shown CPU usage graphs for two typical applications that run on the Pleiades clusters. The profile at the bottom represents the CFD application SpecuLOOS that will be discussed in more detail later, and the profile at the top has been found with a plasma physics particle simulation code. We realize that the CFD program has an average efficiency of around 50%, the Fast Ethernet network is not good enough, whereas the plasma physics application shows an efficiency over 80%, and the Fast Ethernet switch is sufficient. These results are at the origin of the decision to add a GbE switch to Pleiades2, first to satisfy the needs of the CFD codes, and second, due to the higher node performance, the V_c value would have been too high even for plasma physics applications. On Pleiades2 with a GbE, efficiencies for SpecuLOOS are typically over 80%.

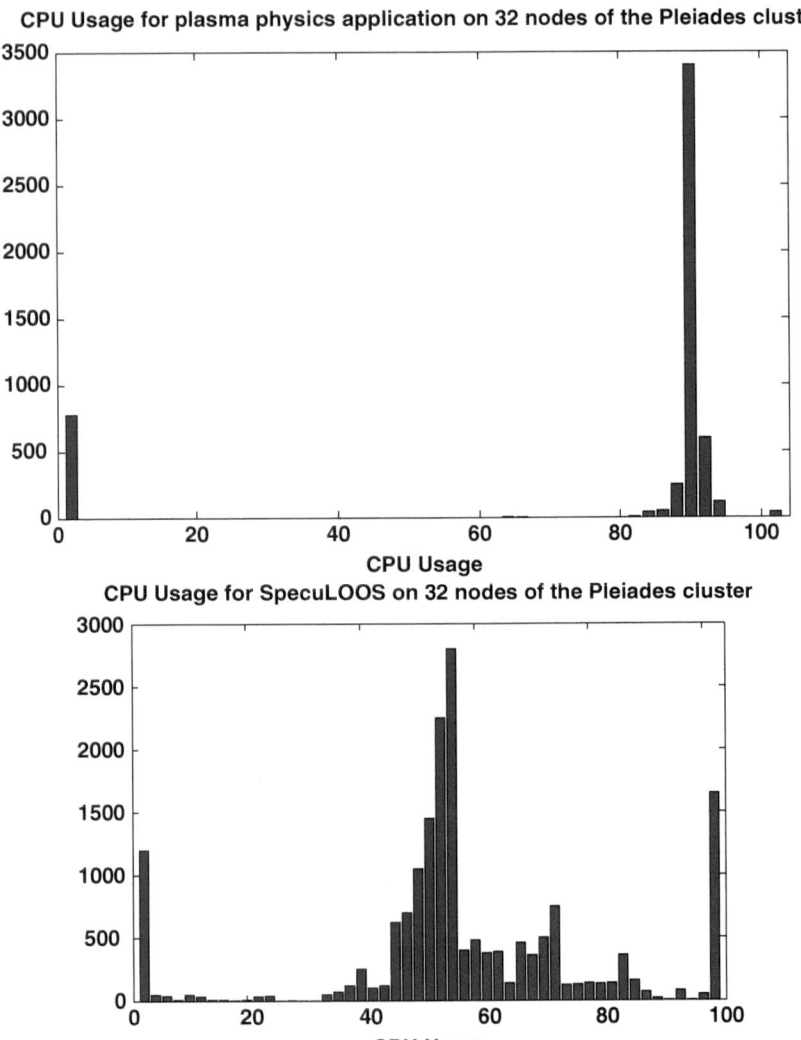

Fig. 7.17 The CPU usage (in %) of two applications in CFD (bottom) and plasma physics (top). The number of time units (y-axis) are collected by Ganglia and attributed to the applications after interpretation of the dayfile

7.3.3 Energy consumption issues

In 2005, 700 millions of Intel processors were installed around the world. If one admits that at a given time 200 millions are active, the power consumption of 200 Watt per processor demands 40 GW sustained, corresponding to 40 1GW nuclear power plants. Here, cooling is an additional energy expenditure. Do we really need high voltage processors, or is pushing up the processor frequency just a commercial argument? In the newest TOP500 list, the energy consumptions are listed. One sees that an IBM/Blue Gene /P core (850 MHz) consumes 7.7 W, 30 W are needed for a recent AMD 2.3 GHz core, and 40 W for a 3 GHz Intel Harpertown core. An older representation of the Performance/Watt can be found in Fig. 2.2. The low energy consumption of the Blue Gene/P made it possible to put 1024 processors (=2048 cores) in a rack. With this machine IBM went in the right direction: Low core frequency; using one core for communication, only one core is needed for small V_a applications; small main memory; when needing more main memory, increase the number of nodes; use different communication networks for different application needs. These ideas should be pushed further.

7.3.4 The issue of resilience

When running on thousands of computational units in parallel, the probability that one node fails becomes an issue. Supercomputing centers often limit the resource usage to $T_{limit} = N_{nodes} * T_{CPUlimit}$ hours for one application. If more resources are needed for one run, the user application has to include a checkpoint/restart facility. After periods of time, data needed to restart the simulation from this checkpoint is written onto disk. Discussions are around since many years to automatically checkpoint all applications that run longer than a defined T_{limit}. For this purpose, everything has to be written on disk, also the huge matrices that in a user defined checkpoint would be reconstructed after the restart.

   Suppose that a node has an average time between failure of T_{MTBF} days, a user runs a job on N_{nodes} nodes during T_{CPU} days, and a checkpoint/restart takes $W_{cp/rs}$ % of the CPU time to write on disk. After how many days, call it $T_{interrupt}$, should a checkpoint be done? The failure probability after $T_{interrupt}$ days is $N_{nodes} * T_{interrupt}/T_{MTBF}$ and the relative time loss due to a checkpoint/restart is given by $W_{cp/rs} * T_{CPU}/T_{crit}$, where T_{crit} is the critical application CPU time at which the checkpoint should be done, i.e. $T_{crit} = T_{interrupt}$. As a consequence, it is encouraged to set a checkpoint/restart after

$$T_{interrupt} = \sqrt{\frac{W_{cp/rs} * T_{CPU} * T_{MTBF}}{N_{nodes}}}. \tag{7.2}$$

Users often complain that after a checkpoint, the job has to be put again in the job queue, resulting in additional waiting time. This problem can be solved by increasing the time limit in the input queue for those codes that have checkpoints. After a checkpoint, the job continues to run. Restart is then only needed if an interrupt really occurs, or if the upper time limit is reached.

7.4 Type of parallel applications

There are different types of parallel applications with different communication needs:

7.4.1 Embarrassingly parallel applications

These applications do not demand inter-node communications. A master node distributes a large number of cases among many slave nodes, it collects and integrates the results. No data is exchanged between slave nodes. In this case, $T_{comp} >> T_{comm}$, and thus $\Gamma >> 1$. As a consequence, very high γ_m communication networks such as a bus, a cluster with a Fast Ethernet or a Gigabit Ethernet, or for cases with a small amount of input and output data, even the Internet can be used. A typical example is the $seti@home$ project that collects computational cycles over the Internet. Other examples of such applications are the immense amount of independent data in high energy physics that has to be interpreted, the sequencing algorithms in proteomics, parameter studies in plasma physics to predict optimal magnetic fusion configurations ([131]), a huge number of data base accesses as Google does, or visualization algorithms.

7.4.2 Applications with point-to-point communications

Point-to-point communications typically appear in finite element or finite volume methods when a huge 3D domain is decomposed in subdomains [67] and an explicit time stepping method or an iterative matrix solver is applied. If one admits that one subdomain is taken care of by one virtual processor, the number of processors grows with the problem size, and the size of a subdomain is fixed, γ_a is constant. The amount of data to be sent and the number of messages that are sent from a virtual processor to the neighboring ones is fixed. Since the sparse matrix times vector is the dominant operation, the per processor performance is determined by the main memory bandwidth and not by the processor speed. The number O of operations per step is directly related to the number of variables in a subdomain times the number of operations per variable, whereas the amount of data S transferred to the neighboring subdomains is directly related to the number of variables on the subdomain surface, and $\gamma_a = O/S$ becomes big. For huge point-to-point applications using many processing nodes, $\Gamma << 1$ for a bus, $2 < \Gamma < 10$ for a Fast Ethernet switch, $10 < \Gamma < 50$ for GbE, Myrinet and Infinibands, and $\Gamma >> 100$ for Cray XT3. Hence, that kind of applications can run well on a cluster with a Fast Ethernet or a GbE switch, but a bus or the Internet are definitely not adequate.

7.4.3 Applications with multicast communication needs

The parallel 3D FFT algorithm is a typical example with important multicast communication needs. Here, γ_a decreases when the problem size is increased, and the communication network has to become faster. In addition, $r_a = R_\infty$ for FFT, γ_m is large, and, as a consequence, the communication parameter b must be large to satisfy $\Gamma > 1$. Another example is a Car-Parrinello molecular dynamics application in material sciences [134]. It has been showed that with a Fast Ethernet based switched network, the communication time is several times bigger than the computing time. It needs a fast switched network, such as Myrinet, Quadrics, Infiniband, or special vendor specific networks such as those of a Cray XT3 or an IBM Blue Gene. A GbE switch can be used, but, however, is affecting the overall performance of an application with multicast communication messages.

7.4.4 Shared memory applications (OpenMP)

There are a few parallel applications that demand a shared memory computer architecture. The parallelism of the component is expressed with OpenMP. A typical example is the one described in [63]. This implies that a HPC Grid should also include SMP or NUMA nodes that can run OpenMP applications such as the new multi-cores and multi-processors units (Intel Woodcrest, AMD Socket F, and their successors).

7.4.5 Components based applications

A multiphysics application can be separated into components, each component computing a separate physics problem. If inter-component communication is not too demanding, each component can run on a separate machine. This is the reason why we talk about components instead of applications. However, most of the present HPC applications consist of one single component. Examples of component-based applications are shown in Figs. 7.19, 7.22, 7.25, and will be discussed later.

7.5 Domain decomposition techniques

When solving partial differential equations on a two-dimensional (2D), or three-dimensional (3D) domain, a huge number of mesh points can lead to matrix problems with main memory needs far above those offered by one node. In such cases, the domain is cut in N_{sd} subdomains such that each subdomain fits the main memory of a processor. Subdomain decomposition is also used to reduce the overall CPU time. The computational work of a subdomain can be distributed among cores through OpenMP, or the number of subdomains connected through MPI fits the number of cores reserved for the computation. In this case, the computational efforts for each subdomain should be well balanced, as for instance in the SpecuLOOS code, Fig. 7.24. If one subdomain takes more CPU time than the others, those have to wait if communication with this subdomain is needed. The domain decomposition technique is presented through the Gyrotron test example.

7.5.1 Test example: The Gyrotron

The 2D axisymmetric simulation code DAPHNE [130] is used to design a high frequency wave generator, called Gyrotron, illustrated in Fig. 7.18. In a Gyrotron, the electron beam is accelerated by a huge electric field, in Fig. 7.20 76 kV are imposed as boundary conditions, reaching velocities of about half of speed of light. This beam is guided by a magnetic field of around 10 Tesla produced by superconducting coils shown in blue (Fig. 7.18) around which the electrons rotate with a frequency of 100 GHz. At the place of the cavity, or resonator, a standing wave at the same frequency pumps energy out of the beam. This wave is injected into a fusion reactor experiment, such as ITER [84], heating there the electrons. After having passed the cavity, the remaining beam energy has to be destroyed by depositing the electrons on the surface of the Gun collector. The magnetic field coils at the end of the gun open

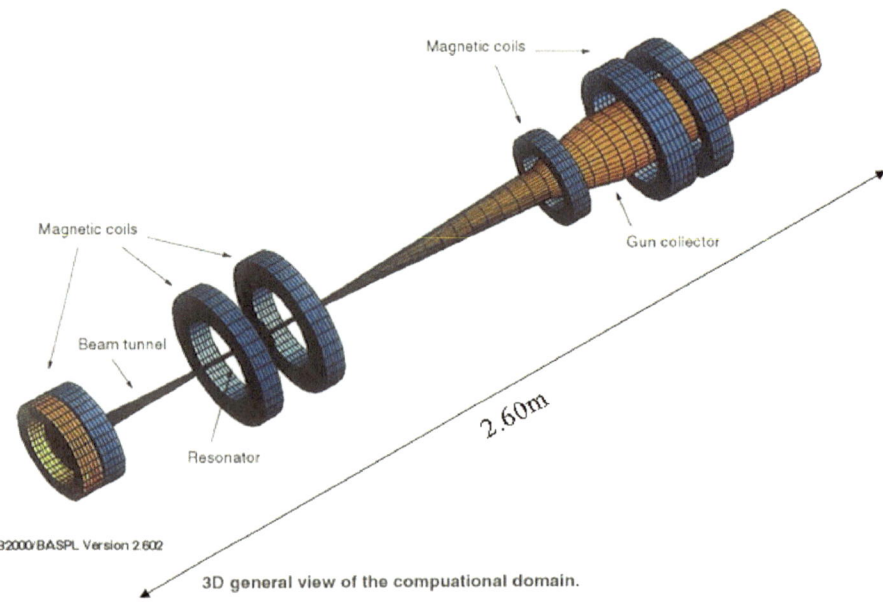

Fig. 7.18 The Gyrotron geometry

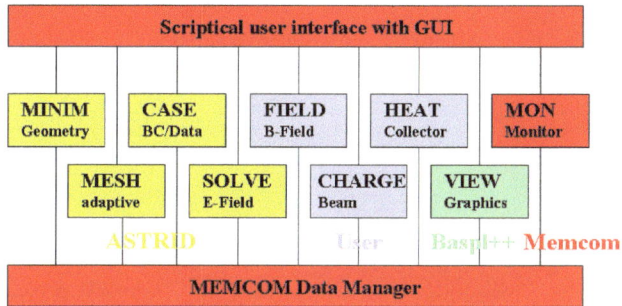

Fig. 7.19 Gyrotron program environment

the beam such that the energy deposition on the collector wall is as homogeneous as possible, thus, not drilling holes in the wall.

The Gyrotron is a multi-physics application realized within the ASTRID [20] program environment. Its organization is shown in Fig. 7.19. The electric field is computed by using finite elements to solve the Poisson equation (SOLVE), the magnetic field is computed by a Biot-Savart method (FIELD). The electron beam is simulated by a number of charged particles ejected from the cathode. The traces of these particles are determined by integrating them ahead in the electromagnetic field using a particle pushing method for relativistic charged particles (CHARGE). The energy deposition on the collector, and the wall cooling are computed in HEAT.

The electric field itself depends on the charge density distribution of the electron beam, Fig. 7.20, needing a non-linear iterative sequence: Compute the electric field with a given charge density distribution, then compute the charge density distribution integrating ahead the electron beam in a fixed electric field, compute again the electric field, and so on. This sequence converges in 3 iteration steps to an electric field that is consistent with the electron beam.

The computational modules are interconnected through the data management system MEMCOM [103]. The overall system is completed by a monitor (MON) that enables accessing data at all time, and by visualization systems (VIEW) using

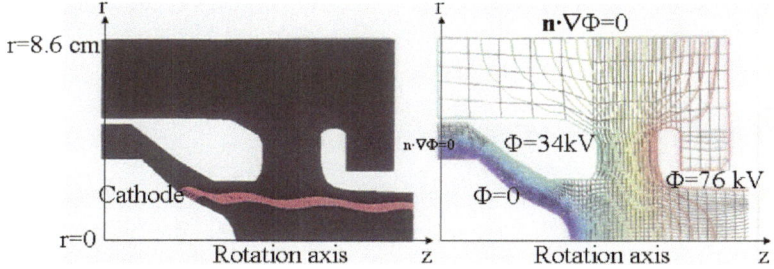

Fig. 7.20 Gyrotron: The electron emitter (left) and the electric field boundary conditions (right), corresponding to the lower left part of Fig. 7.18

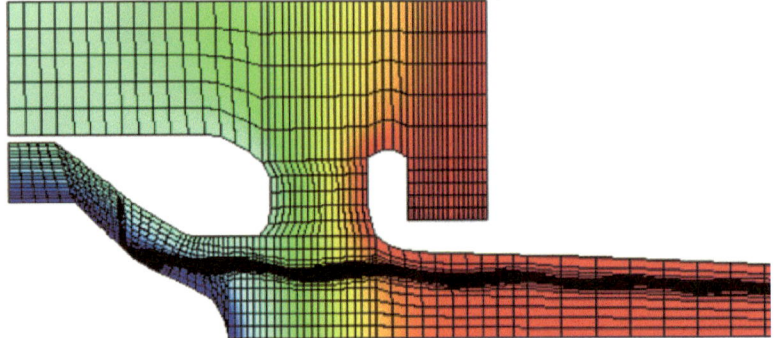

Fig. 7.21 Gyrotron: The beam adapted mesh

BASPL for MEMCOM. The graphical representations in Figs. 7.18, 7.20, 7.21, and 5.17 have been realized with BASPL.

7.5.2 *The geometry and the mesh*

The axi-symmetric Gyrotron geometry is given by a set of CAD points defined by the design engineer. These points are interconnected by straight lines, and by curvilinear sections defined through circles and other curves. This geometrical domain is decomposed in 26 subdomains, each one consists topologically of a deformed square. The subdomains in the emitter part of the Gyrotron can be recognized in Fig. 7.21 that is cut into 14 subdomains. The geometry definition and the subdomain decomposition are handled, i.e. by the module MINIM in Fig. 7.19.

Each subdomain is meshed by the structured adaptive curvilinear mesh generator MESH. The mesh is adapted to the electron paths using the charge density distribution as part of the mesh density function. In the mesh shown in Fig. 7.21 the mesh density function is constant in longitudinal direction, in radial direction the mesh density is given by a constant plus the normalized charge density distribution. This leads to a dense mesh around the beam, with a close to equidistant mesh outside the beam. The mesh density function is continuous across subdomain borders. If the number of mesh cells on the subdomain connecting borders is equal on both sides, the global mesh is continuous everywhere.

7.5.3 *Connectivity conditions*

In 2D, two edges are neighbors if two subdomain corner CAD points are identical. Two corners are neighbors if there is one common CAD point. This determines the geometrical connectivities.

In 3D, two subdomains connect through a surface if the four corner CAD points are identical. If two corner CAD points are identical, the connectivity is across an edge, and if one CAD point is identical, the two subdomains connect through one single corner point.

To ease computation of matrix and vector operations needed in the iterative matrix solver, the master/slave concept is introduced to characterize the border points as to be master or slave points. The choice is made as follows: In subdomain number 1, all the connectivity points are master points. In subdomain number 2, all the common points with subdomain 1 are slave points, all the others are masters. Thus, in subdomain j, all common points with subdomains having numbers smaller than j are slaves, all the others become master points.

7.5.4 Parallel matrix solver

7.5.4.1 Parallel matrix construction

In the ASTRID program environment [20] the matrix is constructed in each subdomain separately in parallel. The submatrices are not assembled. The complete matrix consists of non entirely assembled submatrices in each subdomain plus the connectivity conditions between the subdomains. The matrix problem can be solved by direct (SOLVE in ASTRID), or an iterative solver. In our considerations, we concentrate on a Conjugate Gradient (CG) iterative solving mechanism.

7.5.4.2 Iterative matrix solver

Let us suppose that the full matrix is stored as a sum of non-assembled submatrices for each subdomain, plus the connectivity conditions that link each subdomain to the neighbouring ones. The matrix contributions corresponding to variables at the border of a subdomain are distributed among the different connected subdomains. It is also supposed that all vector components, also those corresponding to a common border point between two neighboring subdomains, are fully assembled before an operation. Specifically, vector components have the same values on all common subdomain border points.

In the conjugate gradient iteration, often used to solve a matrix problem iteratively, there are three parallel operations: the vector product, the dot product, and the sparse matrix * vector multiplication.

The vector product
Since all the vector components are assembled, a vector product

$$\mathbf{y} = \mathbf{y} + a\,\mathbf{x}$$

can be performed in parallel in each subdomain. No communication over the network is needed.

The dot product
In each subdomain, the dot product writes

$$s = \sum_{i=1}^{N_{sd}} \left(\sum_{j=1}^{N_i} x_j * y_j - \sum_{k=1}^{s_i} x_k * y_k \right),$$

where N_{sd} is the number of subdomains, N_i is the total number of mesh points in subdomain i, and s_i is the number of slave points in subdomain i. The dot product can now be performed in three steps:

1. Sum up all the partial sums in each subdomain (j sum from 1 to N_i)
2. Deduct all terms in the dot product that correspond to the slave points (k sum from 1 to s_i). Then, only master points have been added at the borders
3. Perform an *MPlallreduce* operation to sum up all the partial dot products of all subdomains (i sum from 1 to N_{sd}).

Steps 1 and 2 do not imply any communication. In the MPIallreduce operation, the number of messages to be sent over the network is proportional to log (N_{sd}).
The matrix*vector product
This

$$\mathbf{y} = A\,\mathbf{u} = \sum_{i=1}^{N_{sd}} A_i\,\mathbf{u_i}$$

operation has the highest complexity. Here, A_i are the non-assembled submatrices in each subdomain i. The number of operations in each subdomain corresponds to two times the number of non-zero matrix elements. Using MPI-1, this operation is performed in eleven steps:

1. All the $A_i\mathbf{u_i}$ operations are performed in parallel in each subdomain. Note, that compared to a fully assembled matrix, the total number of operations is now slightly bigger
2. The partial results at slave points are assembled in lists
3. The lists are sent to the corresponding master point subdomains. This needs an MPIsend call
4. The master subdomain receives the slave lists calling MPIreceive
5. The slave lists are disassembled
6. The slave components are added to the master components
7. If all the slave components coming from one or more neighbors are added to the master, the updated master components are assembled in lists
8. The lists of master components are sent to the slave subdomains calling MPIsend
9. The master lists are received by the slave subdomain calling MPIreceive
10. The master lists are disassembled
11. The old slave components are replaced by the new master components.

The assembling and disassembling steps are important to reduce the number of messages and the network latency time. In 3D finite element methods, there are maximal 26 neighbouring subdomains, and maximal 26 messages have to be sent and received per data exchange from one subdomain. It has been shown in [20] that this number can be reduced to a minimum of 6 sends and receives per data exchange when introducing ghost subdomains.

The number N_{it} of iteration steps depends on the matrix condition number. To reduce the matrix condition, a preconditioning matrix can multiply A from the left. If this preconditioner is chosen such that it represents a good approximation of the inverse of A, the preconditioned matrix is close to identity, and the CG iteration converges more rapidly. These preconditioners can become quite complex (see for instance [49]).

7.5.5 *The electrostatic precipitator*

Another example, with a much more complex geometry, is the 3D Electrofilter [47], called ESP (Electro-Static Precipitator). This installation can reach the size of a soccer ground, and is typically 15 m high. It is used to eliminate pollutants from the air coming out of an electric power station using coal or oil. The polluted air is injected in the ESP. The air is ionized in the corona region of electrodes set on high voltage of about 35 kV. These charged air particles then move along the electric field to the wall set on earth. On the path towards the wall, ionized air particles enter into collision with the pollutants, ionizing them. The electric field pushes the ionized pollutants to the wall, where they are deposited. Periodically, the pollutants are eliminated from the wall by a vibration process. Electrofilters eliminate over 99% of the pollutants.

The organization of this industrial problem is shown in Fig. 7.22. Again, the geometry and the domain decomposition is done in MINIM. In addition to the previous Gyrotron application, here, the geometry has been parameterized to offer to the user the possibility to modify it easily through 12 parameters. The air flow is described by the Navier-Stokes equations solved in FLOW, the electrostatic field and the charge density distribution are again computed in SOLVE and CHARGE, respectively. The incoming air is ionized in the corona region of the electrode, using a model (CORONA) predicting the ionisation level through the measured normal current at the wall. The particles of the ionised air and of the ionised pollutants are integrated ahead in PATH as in the Gyrotron case.

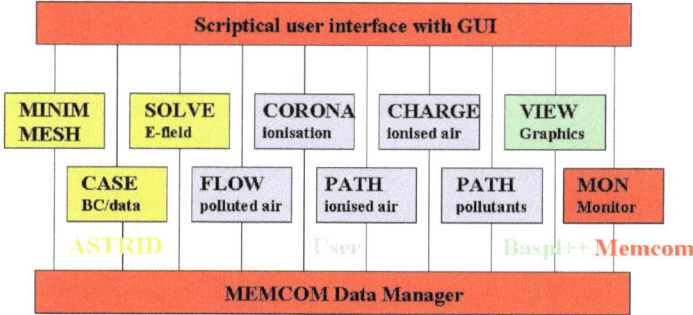

Fig. 7.22 Electrostatic Precipitator: The software environment

7.6 Scheduling of parallel applications

7.6.1 Static scheduling

An application consists of N_{sd} subdomains or tasks. In most parallel applications, a major effort is made to well balance the CPU and the communication network. The reason to do so is that it is believed that one subdomain is taken care of by a node, by a processor, or by a core.

In this case, it is optimal to run the N_{sd} tasks in parallel on $P_i = N_{sd}/k$ computational units that all have the same hardware and basic software architectures. This implies the same CPU, the same memory subsystem, the same main memory size, the same communication network, and the same system and compilers. Often, subdomains at the border of the geometrical domain need less communication than subdomain in the centre (see for instance Fig. 7.24). At barriers, the border subdomains have to wait until the internal subdomains have finished communication. If the number of internal subdomains is much bigger than the number of border subdomains, or if the communication time is smaller than the computation ($\Gamma > 1$), this effect is small. Often, an effort is made to reduce communication time by hiding communication behind computation. If this is possible, the difference between border and internal subdomains can be hidden. But, be aware of possible race conditions.

If a subdomain is taken care of by a core, MPI can be used to activate all computational units, also to internally communicate in SMP or NUMA nodes. If a subdomain is sent to a node or to a processor with several cores, OpenMP can be used for inter-core parallelization, and the application has two levels of parallelization: MPI between nodes, and OpenMP in a node. In older vector machines, the compiler was able to auto-parallelize by distributing the work of multiple DO-loops among the shared memory (SMP) multi processors. The scalability of the resulting code was typically between 50% and 80%. We have seen in the TERPSICHORE case that OpenMP has to be used with care. Often, to get high efficiency, the code optimization has to be made by hand.

7.6.2 Dynamic scheduling

Suppose that an application consists of N_{sd} tasks, and there are P_i computational units (nodes, processors, or cores) on the resource i. If the tasks are not well equilibrated, i.e. if the CPU time and/or the network communication take very different times in different tasks, smaller tasks have to wait results from bigger tasks. The same problem appears if the tasks are well equilibrated, but the computational units and/or the communication networks are not homogeneous, thus take more or less time to execute a task. Then, a dynamic scheduling is perhaps more adequate.

The different tasks of an application, or of a component, are put into a list, according to CPU and communication times evaluated by a complexity modelanalysis.

The task taking most time (CPU plus communication) is at the top, the one with the smallest time is at the end of the list. The fastest computational unit (a node, processor, or a core) takes the first task, the second the next one, until all the units are active. If a computational unit finishes, it takes the next task in the list.

7.7 SpecuLOOS

This section is based on the publication [22].

7.7.1 Introduction

The **Spec**tral **u**nstructured **E**lements **O**bject-**O**riented **S**ystem (SpecuLOOS) is a toolbox written in C++. SpecuLOOS is a spectral and mortar element analysis toolbox for the numerical solution of partial differential equations, and more particularly for solving incompressible unsteady fluid flow problems [133]. The main architecture choices and the parallel implementation were elaborated and implemented in [46]. Subsequently, SpecuLOOS' C++ code has been developed, see [23, 24, 25].

It is well known that spectral element methods are easily amenable to parallelization as they are intrinsically a natural way of decomposing a geometrical domain ([50], and Chap. 8 of [43]).

The references previously given and the ongoing simulations based on SpecuLOOS highlight the achieved versatility and flexibility of this C++ toolbox. Nevertheless, ten years have passed between the first version of SpecuLOOS' code and the present time, and tremendous changes have occurred at both hardware and software levels.

Here we discuss the adaptation of SpecuLOOS to thousands of multi-core nodes. Performance measurements on one-core node, on a cluster with hundreds of nodes, and on a 4096 dual-core Blue Gene/L are presented. The obtained complexities are compared with theoretical predictions. First results show that small cases gave good complexities, but huge cases gave poor efficiencies. These results led to the detection of a poor parallel implementation. Once corrected, the speedup of SpecuLOOS corresponds to the theoretical one up to the 8192 cores used.

7.7.2 Test case description

The test case belongs to the field of CFD and consists in solving the 3D Navier–Stokes equations for a viscous Newtonian incompressible fluid. Based on the problem at hand, it is always physically rewarding to non-dimensionalize the governing Navier–Stokes equations which take the following general form

$$\partial \mathbf{u} + \mathbf{u} \cdot \nabla \mathbf{u} = -\nabla P + \frac{1}{\mathrm{Re}} \Delta \mathbf{u} + \mathbf{f}, \qquad \forall (\mathbf{x}, t) \in \Omega \times I, \qquad (7.3)$$

$$\nabla \cdot \mathbf{u} = 0, \qquad \forall (\mathbf{x}, t) \in \Omega \times I, \qquad (7.4)$$

where **u** is the velocity field, P the reduced pressure (normalized by the constant fluid density), **f** the body force per unit mass, and Re the Reynolds number

$$\text{Re} = \frac{UL}{\nu}, \tag{7.5}$$

expressed in terms of the characteristic length L, the characteristic velocity U, and the constant kinematic viscosity ν. The system evolution is studied in the time interval $I = [t_0, T]$. Considering particular flows, the governing Navier–Stokes equations (7.3)-(7.4) are supplemented with appropriate boundary conditions for the fluid velocity **u** and/or for the local stress at the boundary. For time-dependent problems, a given divergence-free velocity field is required as initial condition in the internal fluid domain.

The test case corresponds to the fully three-dimensional simulation of the flow enclosed in a lid-driven cubical cavity at the Reynolds number of Re $= 12\,000$ placing us in the locally-turbulent regime. It corresponds to the case denoted under-resolved DNS (UDNS) in [25], where the reader can find full details on the numerical method and on the parameters used. The velocity field is approximated in a polynomial space of degree p, and the pressure in a space of degree p-2.

The complexity is proportional to the total number of elements $N = N_x * N_y * N_z$ in the three dimensional space. Each element is transformed to a cube. Since the Gauss–Lobatto–Legendre basis functions

$$h_j(r) = -\frac{1}{p(p+1)} \frac{1}{L_p(\xi_j)} \frac{(1-r^2) L_p'(r)}{(r-\xi_j)}, \qquad -1 \le r \le +1, \quad 0 \le j \le p,$$
$$\tag{7.6}$$

of degrees $p = p_x = p_y = p_z$ are orthonormal, the complexity for the pressure is $(p-1)^3$, while the complexity for the velocity is $(p+1)^3$, and the pressure derivatives have complexities of $(p-1)^4$. During the computations, the variables are frequently re-interpolated between the collocation grids, and the leading complexity of p^4 is due to the tensorization of the implied linear operations. At large values of p, these re-interpolations dominate the total computation time. It should be remarked that from a complexity standpoint, a term like $(p-1)^3$ is equivalent to a term like $(p+1)^3$. In the following, a term $p-1$ has been applied systematically to read the complexity from experimental performance curves, while the notation of the equations is simplified by use of the term p.

The CPU time T of the pressure equation in the SpecuLOOS spectral code is, Eq. (4.1) with $N_2 = 1$,

$$T(N_1, N_{CG}, N, p) = a_1 N_1 N_{CG} N p^{a_3}, \tag{7.7}$$

where N_{CG} is the number of conjugate gradient steps, N_1 is the number of time steps, and a_1 is a parameter that depends on the computer architecture. The quantities N and p are SpecuLOOS input parameters. In Eq. 7.7, the complexity is linear with respect to N, and a_3 is related to the complexity of the underlying algorithm with respect to the polynomial degree p. Specifically, for the matrix solver used in SpecuLOOS, $3 < a_3 < 4$. Both parameters can be found by running at least two cases that have different polynomial degrees p.

7.7.3 Complexity on one node

First, we run SpecuLOOS on one core without any communication using the Couzy preconditioner [37]. One time iteration step of SpecuLOOS is divided into three main parts: (1) computes the tentative velocity (through a Helmholtz equation solved by preconditioned conjugate gradient method), (2) computes the pressure (through a sophisticated conjugate gradient), and (3) corrects the tentative velocity. The relative importance of each of these three components depends on the parameters in a given simulation. With a fine resolution in time, the second part becomes dominant, and takes as much as 90 % of the total CPU time in the example described below.

Table 7.2 presents the results of SpecuLOOS on one node of an Intel Xeon cluster. The CPU time measurements, $T_{CG,(1\ iter)}$, for one time step and one CG iteration step ($N_1 = N_{CG} = 1$) are used to minimize $\sum (T - T_{CG,(1\ iter)})^2$, where

$$T(1, 1, N, p) = a_1 N^{a_2} p^{a_3} . \tag{7.8}$$

With the cases presented in Table 7.2 his minimization procedure leads to the scaling law

$$T(1, 1, N, p) = 2.01 \cdot 10^{-6} N^{0.97} \cdot p^{3.3} . \tag{7.9}$$

One realizes that this complexity law corresponds well to the theoretical one, Eq. (7.7).

Generally, the number of iteration steps is not known. If in the optimization procedure one includes N_{CG} in the parameters N and p,

$$T(N_1, \cdot, N, p) = 1.15 \cdot 10^{-5} \cdot N_1 \cdot N^{1.30} \cdot p^{4.19} . \tag{7.10}$$

Table 7.2 SpecuLOOS on one node of an Intel Xeon cluster. The number of conjugate gradient iterations N_{CG} is an average value over all time steps for the pressure

N_1	N	p	T_{exec} [s]	N_{CG} # iter	T_{CG} [s]	$\frac{T_{CG}}{T_{exec}}$	$T_{CG,(1\ iter)}$ [s]
1	256	8	40.1	198	32.8	0.818	0.17
1	256	10	119.3	247	103.8	0.870	0.42
1	512	6	43.2	205	33.9	0.785	0.17
1	512	8	116.4	268	106.7	0.917	0.40
1	512	10	394.3	344	342.3	0.868	1.00
1	1024	6	105.2	259	83.4	0.793	0.32
1	1024	8	311.0	339	265.4	0.853	0.78

As a consequence, the estimated number of $N_{CG,est}$ is

$$N_{CG,est} = 5.72 \cdot N^{0.33} \cdot p^{0.89} . \tag{7.11}$$

This prediction is also close to the expected theoretical complexity of a CG iteration with a diagonal preconditioner:

$$N_{CG,theo} \approx N^{\frac{1}{3}} \cdot p . \tag{7.12}$$

The same type of studies have been made for a diagonal preconditioner by varying the polynomial degree. The complexity found is that $N_{CG,est} \approx p^{1.47}$. For $p \geq 12$, the diagonal preconditioner is faster, for $p < 12$ the Couzy preconditioner is faster. Since we treat cases with $p = 12$ or smaller, we concentrate on Couzy's preconditioner.

7.7.4 Wrong complexity on the Blue Gene/L

The SpecuLOOS code has been ported to the IBM Blue Gene/L machine at EPFL with 4096 dual core nodes. Since interelement communication is not that important, all the cores can be used for computation. The results are presented in Table 7.3. One element is running per core. The polynomial degree is fixed to $p=12$ for the velocity components, and to $p - 2=10$ for the pressure. The resulting complexity given by the pressure computation is now

$$T \approx N^2 . \tag{7.13}$$

This result shows that the complexity of the original parallel code is far away from the $N^{1.3}$ law, which is expected for theoretical reasons and verified numerically in a serial program execution, Eq. (7.10). The reasons for this bad result could be identified as follows. To identify the attribution of elements to computational nodes dynamically, at each iteration step of the conjugate gradient method an IF instruction over all elements had been introduced in the code. Such an instruction is typical for rapid corrections in a code, which are made to parallelize a program rapidly without realizing its impact on future executions on computers with much more nodes. This did not affect the CPU time for less than $N_{core} = 100$ cores, but became dominant

Table 7.3 SpecuLOOS on the Blue Gene/L machine up to $N_{core} = 8192$ cores. The number of elements per core have been fixed to one. The polynomial degree for the pressure is equal to $p - 2$ = 10

N_1	$N = N_x N_y N_z$	p	N_{CPU}	$\frac{\text{\# elem}}{\text{node}}$	T_{exec}
1	$8 \times 8 \times 16$	12	1024	1	17.22
1	$8 \times 16 \times 16$	12	2048	1	29.91
1	$16 \times 16 \times 16$	12	4096	1	57.05
1	$16 \times 16 \times 32$	12	8192	1	140.50

for $N_{core} > 1000$. In fact, the remark "Valid until 64 processors" could be found in the code. This IF instruction has now been replaced by the use of a pre-computed list, pointing to the elements which are active on a core.

7.7.5 Fine results on the Blue Gene/L

The corrected SpecuLOOS code has been executed again on the Blue Gene/L machine. The effect of the communication has been studied and presented in Fig. 7.23. The straight line represents the ideal speedup. The measurements for the case of one element per core are presented as thick points. Inter-element communication takes a very small amount of time in the case of 1024 elements, and about 12% of the total time for 8192 cores. This good behavior is due to a few big messages that are sent from each node, Fig. 7.24. In addition, the message sizes do not vary much, leading to a well equilibrated communication pattern. If the number of elements per core is increased, the communication to computation ratio decreases, and the computation is closer to ideal scalability.

The measurements in Table 7.4 show that the symmetric cases with $N = 16 \times 16 \times 16$ take less iteration steps than cases with different interval numbers in the different directions, the asymmetric cases. The complexity laws $T \approx N^a$ for the asymmetric cases give exponents of $a_2 = (1.35, 1.28, 1.36, 1.34)$ for $N_{core} = (1024, 2048, 4096, 8192)$. These exponents correspond very well to the expected one, i.e. to $a_2 = 1.3$, Eqs. (7.8), (7.10). This tells us that the present version of the SpecuLOOS code is well scalable on the Blue Gene up to 8192 cores, and will for larger cases scale up to petaflop machines.

7.7.6 Conclusions

The performance review for the high-order spectral and mortar element method C++ toolbox, SpecuLOOS, has shown that its MPI version achieves good per-

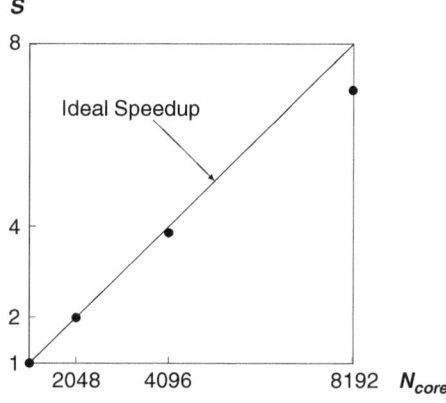

Fig. 7.23 Speedup S on Blue Gene as a function of N_{core} for the case $N = 16 \times 16 \times 32$, Table 7.4. The 12% loss in speedup for $N_{core} = 8192$ is due to communication between the cores. This communication is large since there is only one element per core

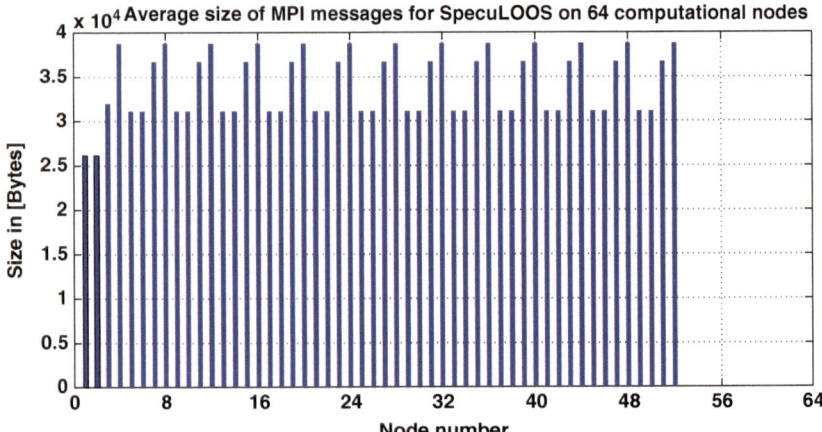

Fig. 7.24 Average size S of the MPI messages for $N_1 = 1$, $N = 512$ and $p = 7$ (for pressure P) on Pleiades2. The number of nodes $N_{nodes} = 64$

Table 7.4 SpecuLOOS on the Blue Gene/L machine up to $N_{core} = 8192$ cores. The polynomial degree for the pressure is equal to $p - 2 = 10$. The number of elements per core varies from 1 to 8. The number of iteration steps is an average value, with a clear drop when $N_x = N_y = N_z$

N_{core}	$N = N_x N_y N_z$	$\dfrac{N}{N_{core}}$	N_{it}	T[sec]
1024	$8 \times 8 \times 16$	1	689	29.9
1024	$8 \times 16 \times 16$	2	966	82.8
1024	$16 \times 16 \times 16$	4	912	155.3
1024	$16 \times 16 \times 32$	8	1463	494.2
2048	$8 \times 16 \times 16$	1	950	42.1
2048	$16 \times 16 \times 16$	2	912	79.1
2048	$16 \times 16 \times 32$	4	1461	249.9
4096	$16 \times 16 \times 16$	1	912	41.9
4096	$16 \times 16 \times 32$	2	1463	128.4
4096	$16 \times 32 \times 32$	4	1925	331.2
8192	$16 \times 16 \times 32$	1	1463	70.7
8192	$16 \times 32 \times 32$	2	1958	179.1

formances on commodity clusters, even with relatively common internode network communication systems, available software and hardware resources.

One goal of this study was to estimate if SpecuLOOS could run on a massively parallel computer architecture comprising thousands of computational units, specifically on the IBM Blue Gene machine at EPFL with 4'096 dual processor units. After detection and correction of a poor implementation choice in the parallel version, perfect scalabilities on up to 8192 cores have been obtained. The present version of the SpecuLOOS code is well scalable above 8192 cores, and will scale up to petaflop machines.

7.8 TERPSICHORE

Let us come back to the MHD stability code TERPSICHORE [65] discussed in chapters 5 and 6. The program includes different components that are represented in Fig. 7.25. After having computed the ideal MHD equilibrium solution with the VMEC [76] code, an equilibrium reconstruction has to be made to construct magnetic flux surfaces that become the new "radial" coordinate. This step is the one taking most of the computing time and demands a big main memory. Then, the matrix is entirely constructed and the lowest eigenvalue computed. At the end, the result is diagnosed and the eigensolution is visualized.

   In this iterative procedure, each equilibrium solution requires several stability computations to be performed with different radial resolutions to be able to predict if an equilibrium is stable or not. This procedure can be done in parallel. Then, the time consuming reconstruction step can be performed in parallel for each radial surface, not needing any communication. Then, in one node, the submatrices are collected and merged to the final matrix. With a BABE (Burn At Both Ends) [62] eigenvalue solver, described in the following section, two nodes can be used to compute the lowest eigenvalue. When all radial resolutions are done, a study can be made on the collected lowest eigenvalues to predict if the converged eigenvalue is stable or not. Independently of all those computations, the diagnostic of the results can be made interactively on a workstation.

   This is an excellent example of a component-wise work-flow that can be executed on a set of resources that can be a cluster, or a GRID of parallel machines and workstations. Each component can be run in parallel. but can profit up to a different number of nodes. The execution steps are governed by the Graphical User Interface in red. Data exchange between the components is made through the Data Manager, also in red.

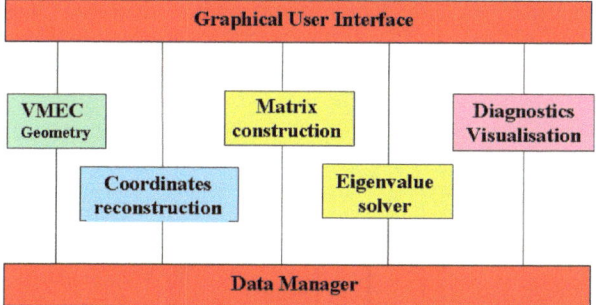

Fig. 7.25 The component-wise structure of TERPSICHORE. The green component is the VMEC 3D MHD equilibrium program. The blue component is the embarassingly parallel coordinate reconstruction taking most of the computing time and most of main memory that runs in parallel. The yellow Eigenvalue solver component is executed using up to two nodes, and the violet Diagnostic/Visualization component is interactively executed on a workstation

7.9 Parallelization of the LEMan code with MPI and OpenMP

7.9.1 Introduction

Thermonuclear fusion is based on the fusion of light nuclei that needs specific conditions to happen. These fusion reactions can be reached for example by confining a very hot, totally ionized gas, called plasma, in a sufficiently strong magnetic field to reach sufficient plasma density and temperature. A great variety of waves can propagate in experimental fusion reactors. The low-frequency domain studied by the LEMan code [109] is especially interesting. Waves can be sources of instabilities or be used for heating purpose to obtain parameters required for a self-sustained reaction.

The concept of the LEMan code is to provide a fast computation of the wave field in fully three-dimensional geometries. As plasma is a charged fluid, it consists then essentially in the direct solution of the Maxwell's equations. These are solved using a Galerkin weak form with a discretization that is characteristic of a toroidal topology: radial finite elements, toroidal and poloidal Fourier harmonics. Such a scheme leads to a full block tri-diagonal matrix for the linear system.

7.9.2 Parallelization

The parallelization of the LEMan code can mainly be separated in three parts: the matrix construction, the solver, and the diagnostics. The third point corresponds essentially to the computation of the plasma quantities relative to the solution (electric and magnetic fields, power deposition, etc). The tasks are however easy to share between processors in this case. We specially concentrate on the two first steps of the computation.

The matrix structure has a block tri-diagonal structure shown below. As the matrix rank is large, the temporary arrays reach quickly the memory limit of the machine. The method that will involve parallelization with MPI has to be implemented directly on the matrix blocks. It uses Gauss elimination which is expressed for one processor as:

$$\begin{pmatrix} B_1 & C_1 & 0 & \cdots & & 0 \\ A_1 & B_2 & C_2 & \ddots & & \vdots \\ 0 & A_2 & B_3 & \ddots & & 0 \\ \vdots & \ddots & \ddots & \ddots & & C_{n-1} \\ 0 & \cdots & 0 & A_{n-1} & B_n \end{pmatrix} \begin{pmatrix} f_1 \\ f_2 \\ f_3 \\ \vdots \\ f_n \end{pmatrix} = \begin{pmatrix} d_1 \\ d_2 \\ d_3 \\ \vdots \\ d_n \end{pmatrix} \qquad \begin{aligned} \mathcal{B}_1 &= B_1, \\ \mathcal{B}_i &= B_i - A_{i-1}\mathcal{D}_{i-1}, \\ \mathfrak{d}_i &= d_i - A_{i-1}\mathfrak{e}_{i-1}, \end{aligned} \qquad (7.14)$$

where $\mathfrak{e}_i = \mathcal{B}_i^{-1}\mathfrak{d}_i$ and $\mathcal{D}_i = \mathcal{B}_i^{-1}C_i$. Once the matrix has been factorized, the second step is to perform the backsolve:

$$f_i = \mathfrak{e}_i - \mathfrak{D}_i f_{i+1}, \tag{7.15}$$

We note that with this method, only one block ($\mathfrak{D}_i$) has to be stored for each radial node. Compared to the usual band matrix storage in LAPACK, this represents a gain of 82%. The main memory concern does not, however, come from the total matrix storage as hard drives can be used for this purpose but from the memory required for the blocks that correspond to a single radial position. As the number of Fourier modes can be several hundred for the more complex geometries and cases, the main memory needs can represent altogether more than 10 GB. In such cases, the optimal machines are SMP whose memory is shared over processors. As OpenMP can be used on those computers, the number of MPI tasks can be kept very low. A method that provides a good scaling is Cyclic Reduction. It has however the disadvantage to require 7 in $(N_{core})/3 (= 12$ when $N_{core} = 32)$ times more operation than a simple Gauss decomposition in the present case where the matrix blocks are full. This technique becomes then faster than a serial run only with more than 32 MPI tasks. With the possibility to take advantage of OpenMP and other parallelization methods, the use of Cyclic Reduction can be avoided. In what follows, we will concentrate on the optimization of the solver with a much lower number of processors.

The first technique that is used is a two-processor method. The method called BABE consists in applying a Gauss decomposition with one processor from the top and with the other from the bottom of the matrix:

$$\text{Processor 1: } \mathfrak{B}_i = B_i - A_{i-1}\mathfrak{D}_{i-1}, \quad \text{Processor 2: } \mathfrak{B}_i = B_i - C_i\mathfrak{D}_{i+1},$$
$$\text{where } \mathfrak{D}_i = \mathfrak{B}_i^{-1}C_i, \qquad\qquad \text{where } \mathfrak{D}_i = \mathfrak{B}_i^{-1}A_{i-1},$$
$$\mathfrak{d}_i = d_i - A_{i-1}\mathfrak{e}_{i-1}. \qquad\qquad \mathfrak{d}_i = d_i - C_i\mathfrak{e}_{i+1}. \tag{7.16}$$

The elimination process is performed until a central system that contains 4 blocks is obtained:

$$
\begin{pmatrix}
\mathfrak{B}_1 & C_1 & 0 & & \cdots & & 0 \\
0 & \mathfrak{B}_2 & \ddots & & & & \\
0 & \ddots & \ddots & C_{\frac{n}{2}-1} & \ddots & & \\
& & 0 & \mathfrak{B}_{\frac{n}{2}} & C_{\frac{n}{2}} & & \\
\vdots & & \ddots & A_{\frac{n}{2}} & \mathfrak{B}_{\frac{n}{2}+1} & 0 & \\
& & & A_{\frac{n}{2}+1} & \ddots & \ddots & 0 \\
& & & & \ddots & \mathfrak{B}_{n-1} & 0 \\
0 & & \cdots & & & 0 & A_{n-1} & \mathfrak{B}_n
\end{pmatrix}
\begin{pmatrix}
f_1 \\ f_2 \\ \vdots \\ f_{\frac{n}{2}} \\ f_{\frac{n}{2}+1} \\ \vdots \\ f_{n-1} \\ f_n
\end{pmatrix}
=
\begin{pmatrix}
\mathfrak{d}_1 \\ \mathfrak{d}_2 \\ \vdots \\ \mathfrak{d}_{\frac{n}{2}} \\ \mathfrak{d}_{\frac{n}{2}+1} \\ \vdots \\ \mathfrak{d}_{n-1} \\ \mathfrak{d}_n
\end{pmatrix}. \tag{7.17}
$$

Its solution is given by the following expression:

$$f_{\frac{n}{2}} = \left(\mathfrak{B}_{\frac{n}{2}} - C_{\frac{n}{2}}\mathfrak{B}_{\frac{n}{2}+1}^{-1}A_{\frac{n}{2}}\right)^{-1}\left(\mathfrak{d}_{\frac{n}{2}} - C_{\frac{n}{2}}\mathfrak{B}_{\frac{n}{2}+1}^{-1}\mathfrak{d}_{\frac{n}{2}+1}\right) \tag{7.18}$$

Once this element has been computed, the backsolve can be undertaken simultaneously until the top and bottom of the matrix:

$$\text{Processor 1: } f_i = \mathfrak{e}_i - \mathfrak{D}_i\, f_{i+1}. \quad \text{Processor 2: } f_i = \mathfrak{e}_i - \mathfrak{D}_i\, f_{i-1}. \qquad (7.19)$$

Such a method has the advantage to divide the time by two in separating totally the tasks between processors. As for the simple Gauss decomposition, only the $\mathfrak{D}_i$ block is stored involving a reduced memory usage. It must be mentioned that this element has a different definition for each processor.

In order to reduce the computation time further, other possibilities exist but do not give the same scalability as the BABE algorithm [62]. For example, the computation of the matrix elements and the solver can be alternated. This method gives very different results depending on the resolution used for the problem. It is obvious that the biggest gain is obtained if those two tasks last the same amount of time.

Finally $\mathfrak{D}_{i-1}$ in Eq. (7.14) is computed by factorizing $\mathfrak{B}_{i-1}$ and solving with C_{i-1} as right-hand side. It is possible to take advantage of the fact that in this case the solution needs three times more time than the factorization. All processors perform then the decomposition as the columns of the right-hand side matrix are shared among them. Possible gain with this technique is limited by the time required to undertake the factorization.

7.9.3 CPU time results

In this section, we will concentrate on three types of cases that appear when performing computations with LEMan. It must be mentioned that the requirements depend mostly on the model under consideration and on the geometry. As we want to compute the wave propagation in a plasma, Maxwell equations are used and can be written as:

$$\nabla \times \nabla \times \mathbf{E} - k_0^2 \hat{\epsilon} \cdot \mathbf{E} = i k_0 \frac{4\pi}{c} \mathbf{j}_{ant}. \qquad (7.20)$$

where $\mathbf{E}$ is the electric field and $\mathbf{j}_{ant}$ is the antenna excitation that appears in the right-hand side of the linear system. The ϵ term is the dielectric tensor. As it relates together the electric current density and the electric field, the physical model is crucial to determine its value. In the cold formulation it is calculated with the help of Newton's equation by considering a charged element of fluid submitted to an electromagnetic field. In this case its value is obtained in the real space and can be inserted directly inside the equation to be solved. As the same number of operations is required to compute each term of the matrix, it is proportional to N_{mn}^2 where N_{mn} is the number of Fourier Harmonics. The solver in itself involves inversions and multiplications of square matrices with N_{mn} rows and columns and scales then as N_{mn}^3. With a great number of Fourier harmonics, the solver dominates over the matrix computation.

In the warm model where the effects of the distribution function of the particles in the velocity space have to be taken into account, the requirements for the matrix computation change drastically. The dielectric tensor is then calculated by using the Vlasov equation which describes the distribution function (f) evolution:

$$\frac{\partial f}{\partial t} + \mathbf{v} \cdot \frac{\partial f}{\partial \mathbf{x}} + \frac{q}{m} [\mathbf{E} + \mathbf{v} \times \mathbf{B}] \cdot \frac{\partial f}{\partial \mathbf{v}} = 0. \tag{7.21}$$

Several simplifications are performed on (7.21) postulating that the radius of the particle trajectories around the magnetic field lines is negligible compared to the wavelength of the perturbation and to the characteristic length of variation of the plasma parameters. In order to solve the relation obtained after simplification and to conserve the exact expression for all the terms in general three-dimensional geometry, the dielectric tensor is determined as the convolution connecting together the electric current density and the electric field. The inversion of a polynomial linear system of degree 1 in $v_{\parallel}$ is in this case needed:

$$\begin{pmatrix} a_{1,1} + b_{1,1}v_{\parallel} & a_{1,2} + b_{1,2}v_{\parallel} & \cdots & a_{1,p} + b_{1,p}v_{\parallel} \\ a_{2,1} + b_{2,1}v_{\parallel} & a_{2,2} + b_{2,2}v_{\parallel} & & a_{2,p} + b_{2,p}v_{\parallel} \\ \vdots & & \ddots & \\ a_{p,1} + b_{p,1}v_{\parallel} & a_{p,2} + b_{p,2}v_{\parallel} & & a_{p,p} + b_{p,p}v_{\parallel} \end{pmatrix} \begin{pmatrix} f_{l,1} \\ f_{l,2} \\ \vdots \\ f_{l,p} \end{pmatrix} = g(\mathbf{v}, \mathbf{E}, l). \tag{7.22}$$

In this case, the matrix construction is much longer than the solver.

7.9.3.1 Warm model

The first situation is relative to the warm model. The speed-up plotted against the number of processors is displayed in Fig 7.26. As the computation time is much higher for the matrix construction than for the solver, the parallelization consists simply in sharing equally the number of radial nodes on every processor. The CPU time behavior is then theoretically $1/N_{core}$. Fig 7.26 shows that it is effectively close to it. In the warm case, the parallelization seems not to be a major problem.

7.9.3.2 Cold model

Now that the warm case, where simple parallelization can be performed, has been investigated, we will concentrate on the cold model where it is subtler. Two different situations are presented. The first one contains a reduced number of Fourier harmonics ($N_m = 319$). The idea is that the matrix construction and the solver take almost the same time. As we work on a SMP machine, it is also interesting to compare parallelization between MPI and OpenMP. The results are presented in Figure 7.27 against the ideal behavior. OpenMP has a better scaling than MPI but is far away from perfect when it reaches 16 processors. It must be pointed out than

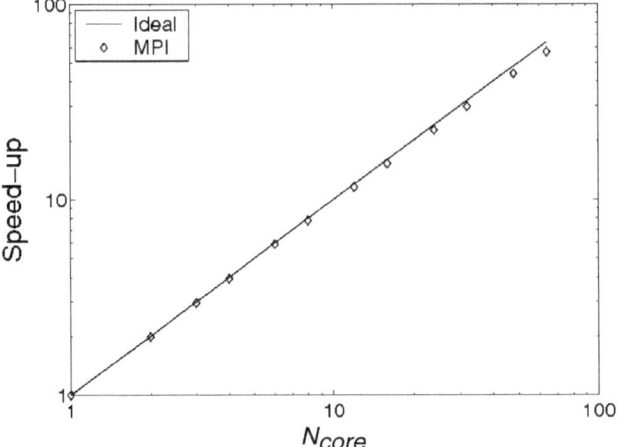

Fig. 7.26 Speed vs number of processors for the warm model with 96 Fourier harmonics

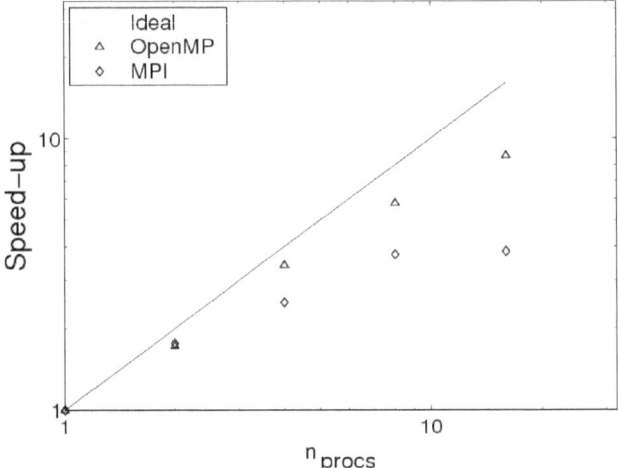

Fig. 7.27 Comparison of the speedup with a parallelization using OpenMP and MPI for the cold model with an intermediate number of Fourier harmonics ($N_m = 319$)

this technique has been implemented in order to compute cases with a high number of Fourier harmonics. This is obviously not the case here. Concerning the behavior of the MPI curve, some explanations must be given. The first scheme used for 2 processors is BABE [62]. The speedup is very close to what has been obtained with OpenMP but is not as perfect as expected. The step to 4 processors is performed by alternating the matrix construction and the solver. This is quite efficient as those two computations take roughly the same order of computational time. For a higher number of tasks, the separation of the right-hand side matrices in the solver between processors has been used. The gain of time is quite interesting from $N_{core} = 4$ to

Fig. 7.28 Comparison of the speedup with a parallelization using OpenMP and MPI for the cold model with a high number of Fourier harmonics ($N_m = 841$)

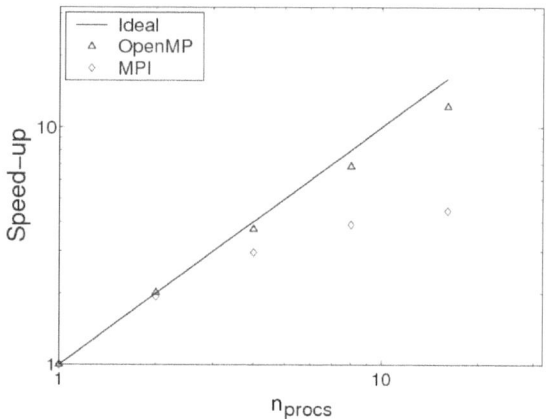

$N_{core} = 8$ but it is practically negligible for the last step. MPI seems then to give satisfactory results when limited to less than 8 tasks.

The final simulation is made with a higher number of Fourier modes ($N_m = 841$). Results are expected to be better as the code is mainly optimized for this kind of situation. They are effectively those which require the most resources. The speedup is shown in Fig. 7.28. Again OpenMP exhibits a better dependence than MPI. In this case, this method uses BABE [62] for two processors. For a higher number of tasks, sharing of the right-hand side matrices has been used for the solver. Going from 2 to 4 processors is efficient with this method. For a higher number of tasks the gain progressively diminishes.

The highest number of Fourier harmonics that has been used with the LEMan code is around 2000. In this case, the dependence in increasing the number of processors is better than for the cases shown here. Taking account of them, we can deduce that the best results would be obtained using the BABE method and OpenMP inside the nodes. In that situation, a speedup of about 24 is reached. A possible gain with cyclic reduction would be achieved with more than 744 processors as this method requires more than 12 times more operations. This is a very large requirement for solving of a single linear system. With 32 processors, the largest case can take about 2 days. Furthermore, this computation applies for a single frequency. For a frequency scan, it is possible to separate the computation by dividing the spectrum.

7.10 Conclusions

We experienced that there are simulation codes that are dominated by internode communications, as in the Helmholtz solver presented in Section 6.3. Here, the FFT algorithm demands matrix transposition operations that are very network dominated. Such a code needs a single-node SMP architecture with sufficient shared main memory.

The SpecuLOOS program is based on the spectral element CFD algorithm. For linear elements, it is equivalent to a finite element approach. If one fixes the problem size per core, the iterative matrix solver is dominated by point-to-point message passing operations, and scales up to a large number of cores. This is an ideal application to run on supercomputers with up to one million of cores.

In the MHD stability analysis made in the TERPSICHORE code to study the stability of the Stellerator 3D fusion devices, different components with different complexities have to be executed. The matrix construction component is embarassingly parallelizable, a large number of nodes can be activated with small internode connection needs. The second component is the eigenvalue solver based on a BABE [62] algorithm with a parallelization level of two.

In the LEMan code there are two different physical models programmed, the cold and the warm plasmas. The warm model for plasma wave propagation, destabilization, and absorption has been shown to give a good scaling with a simple decomposition of the task along the magnetic surfaces. The size of the problem can be increased directly by incrementing the number of processors. For the cold model, the problem is more complex as it depends on the characteristics of the resolution. A high number of Fourier harmonics gives a better scaling. A balance has to be found for the parallelization between MPI and OpenMP. If the SMP nodes contain a sufficient number of processors, the best method is obviously to use two nodes related by the BABE algorithm when the computation is parallelized with OpenMP. If the cyclic reduction would be implemented, a huge number of processors would then be required for a single linear system to be solved.

Chapter 8
Grid-level Brokering to save energy

"We choose to go to the moon. We choose to go to the moon in this decade and do the other things, not because they are easy, but because they are hard, because that goal will serve to organize and measure the best of our energies and skills, because that challenge is one that we are willing to accept, one we are unwilling to postpone, and one which we intend to win, and the others, too."

John F. Kennedy, 35th US President

Abstract This chapter presents the time, cost, and energy saving **I**ntelligent **A**pplicatio**N** **O**riented **S**cheduling (hereafter **ïanos**) solution to adapt a set of computing resources to the needs of the applications. A scheduling scenario is proposed in Section 8.2. In Section 8.3, the Cost Function Model (CFM) is described that minimizes an objective function based on one single metric, the financial cost. Then, the ïanos implementation is presented, and validated with a Grid of HPC clusters.

8.1 About Grid resource brokering

Suppose a community of application users who share a set of computing resources of different types, namely a Grid as defined in chapter 3. The goal of the *Grid-level resource brokering* is to automatically find the r_i that is best fitted to a given set of applications $\mathcal{A}$ at a given moment t.

Grid-level resource brokering is the art to find arguments to choose a suited resource in a Grid to execute an application. These arguments are related to time, funding, policies, and energy. Some users need the result as rapidly as possible, others would like to run at minimum cost, or run the applications as ecologically as possible, and the managers of the computing centers would like to sell as many cycles as possible, and have to follow political rules. These contradictions have to be brought into a unique common theory.

In the ïanos environment, a cost function model has been developed that includes all the upper mentioned arguments. It is based on the parameterization (Chapter 3) of applications and resources, on the Performance Prediction Models (Chapter 4), on resource availability information delivered by a Meta-Scheduler, on resource, infrastructure, software licensing, and energy costs. The data monitoring system VAMOS regularly collects application-relevant quantities during execution and stores them

R. Gruber, V. Keller, *HPC@Green IT*,
DOI 10.1007/978-3-642-01789-6_8, © Springer-Verlag Berlin Heidelberg 2010

on the data base. This data is reused in the cost function model, and also to help managers to decide when which machine should be decommissioned and replaced by what kind of computer architecture. The monitored data can also reveal some application flaws, and help to correct them.

8.2 An Introduction to ïanos

The job submission process in ïanos is decomposed in three main stages:

1. a **prologue**
2. a **cost computation** with the **submission** of the application to the chosen resource
3. an **epilogue**

The prologue

The prologue phase answers the security and availability questions *"is user u* **authorized** *to use the resource r_i?"* and *"is the application component a_k* **available** *on the resource r_i?"*, and checks if the resource can satisfy application needs such as main memory size. This first phase produces a list of **eligible resources** $\mathcal{R}_j$ for the scheduling process of a_k.

The prologue starts at submission time t_k^0.

The cost calculation and application submission

Based on all the relevant quantities that enter an **objective function** computed during this phase, the system produces a list of **configurations** $\mathcal{C}_{k,j}(r_i, a_k, P_j, t_{k,j}^s, t_{k,j}^e)$, $j = 1, \ldots, N_{conf}$ on a set of **eligible resources** $\mathcal{R}_{k,j}$ to execute a_k. Here, r_i is the resource, a_k is the application component, P_j the number of compute nodes reserved by a_k, $t_{k,j}^s$ is the start time and $t_{k,j}^e$ is the predicted end time of execution. The difference $t_{k,j}^e - t_{k,j}^s$ is called **execution time**.

All the configurations $\mathcal{C}_{k,j}$ are classified from the best configuration to the worst based on the QoS demanded by the user. In the present implementation, the five best configurations are sent to the Meta-Scheduling Service (MSS). MSS then chooses the best configuration and submits the application if the resource is still available. Otherwise, the second best is chosen, then the third one, and so on. If all five configurations are no longer available, the system re-starts the process.

During the execution of the application component a_k on P_j nodes of the resource r_i, the performance R_a, the network related data such as number and size of MPI messages, memory usage, or cache misses are monitored and data stored on a data base.

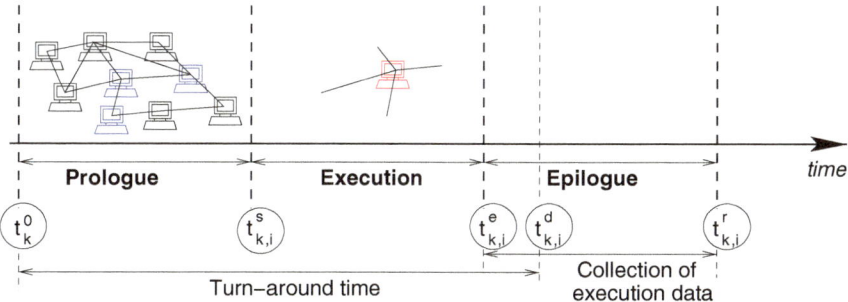

Prologue   **Execution**   **Epilogue**   *time*

t_k^0 $t_{k,i}^s$ $t_{k,i}^e$ $t_{k,i}^d$ $t_{k,i}^r$

Turn–around time

Collection of execution data

Fig. 8.1 ïanos job submission timings for an application component a_k. The difference of timing between $t_{k,j}^e$ and $t_{k,j}^d$ is due to the network and the size of the results that must be sent to the user. Blue resources are $r_i \in \mathcal{R}_k$, the red resource is the one where the application will be submitted

The epilogue

The epilogue starts when the application a finishes. This corresponds to time $t_{k,j}^e$. All the monitored data are stored in an ad-hoc system to be reused for a next submission. The application results are sent to the user. When the user gets all his results, the submission process is over for the user. We are at time $t_{k,j}^d$. The difference $t_{k,j}^d - t_k^0$ is called **turnaround time**. Finally, when all the monitored data are stored in the

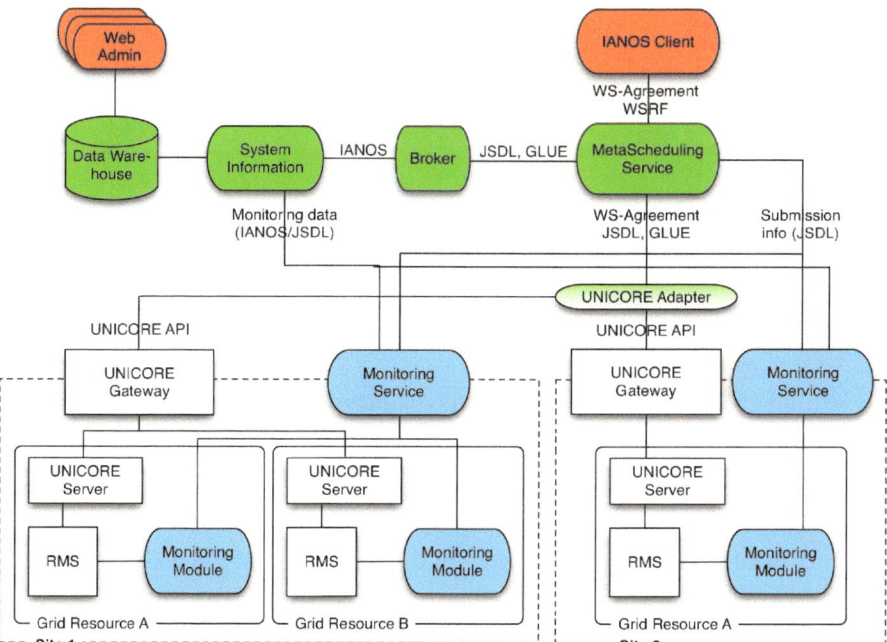

Fig. 8.2 High-level ïanos architecture

storage system (a data-warehouse), the submission process is finished. This time is $t^r_{k,j}$. We have:

$$t^0_k \leq t^s_{k,j} \tag{8.1}$$
$$t^s_{k,j} < t^e_{k,j}$$
$$t^e_{k,j} < t^d_{k,j}$$
$$t^d_{k,j} \leq t^r_{k,j}$$

8.2.1 Job Submission Scenario

We describe here the reference scenario of a job submission using the ïanos framework. SI refers as *System Information*, MSS refers as *Meta-Scheduling Service*, RB refers as *Resource Broker*, MS refers as *Monitoring Service*, MM refers as *Monitoring Module* such as depicted on Figure 8.2.

1   The user u logs into the ïanos client.
2   The user u requests the applications from MSS by sending a WS-Agreement (Web-Service-Agreement which is a protocol that describes an agreement between a web-service and a consumer [2]). It is a WS-Agreement template request.
3   MSS validates and authenticates the client request, contacts the Grid Adapter (the interface between the ïanos framework and the underlying middleware) for all the Grid Sites, and then queries these sites for the list of installed applications based on user authorization filtering.
4   Prepares and returns WS-Agreement templates of the applications, to the client.
5   The user selects one application manually; specifies application intrinsic parameters and the QoS preference (Cost, Time, Optimal), and submits the application as an agreement offer to MSS.
6   MSS validates agreement offer, selects the candidate resources based on the user access rights and the application availability on each candidate resource. It queries the Grid Adapter for each candidate resource static and dynamic information including the candidate resources local Resource Management System (RMS) availability.
7   MSS sends the application's input, the candidate resources with their availabilities, intrinsic parameters, and QoS preference to the RB.
8   RB requests data from SI on candidate resources, and on the submitted application.
9   SI contacts DW through the Data Warehouse interface for information on the candidate resources, on fix application data, and on previous execution data of the same application.
10 SI sends back requested data including the Execution Time Evaluation Model (ETEM) parameters to the RB

11 RB filters out unsuitable candidate resources based on the application require-
    ments (such as number of nodes, memory size, libraries, licensing). It then
    evaluates the cost function and prepares a list of cost function values and tol-
    erances for all candidate resources based on user QoS preference.

12 RB selects an ordered list of suitable execution configurations after applying the
    cost function to all candidate resources, and sends the five best cases to MSS.

13 MSS uses Grid Adapter to schedule one of the configurations following the
    preferences expressed in the ordered list, and WS-Agreement Resource is cre-
    ated for each successfully submitted configuration.

14 MSS sends submission information along with configuration to MS, which in
    turn sends the same submission information to MM.

15 MM monitors the execution of the application, computes the execution relevant
    quantities and sends them to MS.

16 MS sends the monitored data received from MM to System Information.

17 SI computes execution time evaluation quantities from previous execution data
    and application intrinsic parameters to be used for next application submission.

18 At the end of execution, results are sent to the client.

8.3 The cost model

The heart of the Resource Broker is an objective function based on one crite-
rion which is the overall cost of a submission that should be minimized. This
cost includes computing cost (section 8.3.2) on the target resource, licenses cost
(section 8.3.3), data transfer cost (section 8.3.6), ecological cost (section 8.3.5), and
waiting time cost (section 8.3.4). A few free parameters can be tuned by the user.

8.3.1 Mathematical formulation

The choice of a well suited machine depends on user requisites. Some users would
like to obtain the result of their application execution as soon as possible, regardless
of costs, some others would like to obtain results for a given maximum cost, but in
a reasonable time, and some others for a minimum cost, regardless of time. This is
a *Quality of Service* prescribed by the user.

   The cost function z that should be able to satisfy users' QoS depends on costs due
to machine usage, denoted by K_e, due to license fees K_ℓ, due to energy consumption
and cooling K_{eco}, due to waiting results time K_w, and due to data transfer K_d. All
these quantities depend on the application components (a_k), on the per hour resource
costs (K_i), when reserving $P_{k,i}$ computational nodes on resource r_i for component
k. The user can prescribe the two constraints K_{MAX} (maximum cost) and T_{MAX}
(maximum turn around time). The optimization problem then writes:

$$\min z = \beta K_w \left(\bigcup_{k=1}^{n} (a_k, r_i, P_{k,j}) \right) + \sum_{k=1}^{n} \mathcal{F}_{a_k}(r_i, P_{k,j}), \forall \, \ell \leq k \leq n$$

$$\text{such that } \sum_{k=1}^{n} \Big(K_e(a_k, r_i, P_{k,j}, \varphi) + K_\ell(a_k, r_i, P_{k,j}) \tag{8.2}$$

$$+ K_{eco}(a_k, r_i, P_{k,j}) + K_d(a_k, r_i, P_{k,j}) \Big) \leq K_{MAX}$$

$$\max(t_{k,i}^{d}) - \min(t_k^0) \leq T_{MAX}$$

$$(r_i, P_{k,j}) \in \mathcal{R}(k, j),$$

where

$$\mathcal{F}_{a_k}(r_i, P_{k,j}) = \alpha_k \Big(K_e(a_k, r_i, P_{k,j}, \varphi) + K_\ell(a_k, r_i, P_{k,j}) \Big)$$

$$+ \gamma_k \Big(K_{eco}(a_k, r_i, P_{k,j}) \Big) \tag{8.3}$$

$$+ \delta_k \Big(K_d(a_k, r_i, P_{k,j}) \Big) \quad [\text{ECU}],$$

$$\alpha_k, \ \beta, \ \gamma_k, \ \delta_k \geq 0, \tag{8.4}$$

$$\alpha_k + \beta + \gamma_k + \delta_k > 0, \tag{8.5}$$

and $\mathcal{R}(k, j), \ j = 1, \ldots, N_{conf}$ is the eligible machine configurations for application's component a_k. All the terms are expressed in Electronic Cost Units ([ECU]) that can be chosen by the manager of a resource. The quantities t_k^0 and $t_{k,j}^d$ represent the job submission time and the time when the user gets the result, respectively (see Fig. 8.1).

In our model, the parameters α_k, β, γ_k, and δ_k are used to weight the different terms. They can be fixed by the users, by the computing centre, and/or by a simulator, and represent the mathematical formulation of the QoS. For instance, by fixing $\alpha_k = \gamma_k = \delta_k = 0$ and $\beta \neq 0$, one can get the result as rapidly as possible, independent of cost. By fixing $\beta = 0$ and $\alpha_k, \gamma_k, \delta_k \neq 0$, one can get the result for minimum cost, independent of time. These four parameters have to be tuned according to the policies of the computing centers and user's demands. In the case of the Swiss Grid Initiative SwiNG[1], the overall usage of the machines should be as high as possible. For instance, by adequate choices of $\alpha_k \neq 0$ and $\beta \neq 0$, it is possible to optimally load the resources, and increase usage of underused machines. One recognizes that a simulator is needed to estimate these parameters.

In fact, the user's (resource consumer) and the computing center's (resource provider) interests are complementary, the first ones would like to get a result as soon as possible and for the smallest costs, and the second ones would like to get highest profit. The simulator will be used to try to satisfy both. This implies a constant tuning of the free parameters.

---

[1] http://www.swing-grid.ch/

8.3.2 CPU costs K_e

$$K_e(a_k, r_i, P_{k,j}, \varphi) = \int_{t_{k,j}^s}^{t_{k,j}^e} k_e(a_k, r_i, P_{k,j}, \varphi, t)\, dt \ [\text{ECU}]. \tag{8.6}$$

Each computing center has its specific accounting policy, but often they just bill the number of CPU hours used. Figure 8.3 shows an example of $k_e(t)$ when day time, night time and weekends have different CPU costs.

The CPU costs include the investment S_c^i made at the start of the service period T_0^i, the maintenance fees S_m^i, the interests S_b^i that have to be paid to the bank, the personnel costs S_p^i, the infrastructure S_I^i including the building, the electricity and cooling installations, and the energy costs due to cooling, the management S_a^i overhead, the insurance fees S_f^i, and the margin S_g^i. If a real bill is sent to a user, sales tax has to be included. Presently, the costs for CPU time, data storage and archiving are not separated. In future, a special (costly) effort has to be made to guarantee data security. Note, that the energy costs E_h^i per hour and node are taken care of by a separate term in the cost function (see section 8.3.5).

The price/performance ratio of the most recent machines appearing on the market reduces typically by a factor of close to 2 every year. This implies that the investment S_c^i should enter the CPU costs in a non-linear manner. It is reasonable to define a regression curve $\rho(T, r_i)$ for each machine in the Grid that measures the depreciation of the resource as a function of time

$$\rho(T, r_i) = \frac{S_c^i \rho_i \ln(y_i)}{1 - y_i^{-\rho_i T_i}} y_i^{-\rho_i T} \tag{8.7}$$

with

$$\int_{T_0^i}^{T_0^i + T_i} \rho(T, r_i)\, dT = S_c^i \tag{8.8}$$

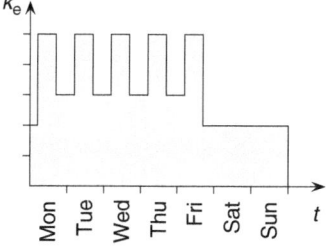

Fig. 8.3 Example of CPU costs as a function of daytime: During daytime the cost is higher, during the night and the week-end lower

that takes this fact into account. The machine installation date of resource i is T_0^i, the life time in years of a machine is T_i, $y_i = 2$, $\rho_i = 1$, and $T_i = 3$ implies that the value of a machine reduces by a factor of 2 every year, and that the machine will be closed after 3 years. These parameters can be chosen by the computing centre.

To compute the CPU costs of a_k it is supposed that $k_e = 1$, not changing during the week time. Admitting that the machine with P_i nodes runs with an efficiency of $e_i\%$ over the year ($d = 8760$ hours/year or $d = 8784$ hours/year in the case of a leap year), the CPU cost K_e of a_k ($P_{k,j}$ nodes, execution starts at $t_{k,j}^s$, and ends at $t_{k,j}^e$) is

$$K_e(a_k, r_i, P_{k,j}) = P_{k,j} \left[\frac{S_c^i}{1 - y_i^{-\rho_i T_i}} \cdot \left(y_i^{-\frac{\rho_i}{de_i}(t_{k,j}^s - t_k^0)} - y_i^{-\frac{r_i}{de_i}(t_e^k - t_0^i)} \right) + S_i(t_e^k - t_s^k) \right]$$

(8.9)

where

$$S_i = (S_m^i + S_b^i + S_p^i + S_I^i + S_a^i + S_f^i + S_g^i)/(de_i P_i).$$ (8.10)

The new quantity S_i denotes the fixed costs per CPU hour for one node, and T_0^i is the age of machine i in hours. With the normalization of ρ_i by de_i, the times $t_{k,j}^s$ and $t_{k,j}^e$ are measured in hours (upper case times are in years, lower case times are in hours). All those values can be given by the computing center through a GUI, Table 8.1. With the ïanos model and a simulator, we hope that it will be possible to estimate T_i, i.e. the time at which a machine should be replaced by a more recent one.

The φ parameter introduces the **priority** notion [64]. Some computing centers do not permit priority ($\varphi = 1$ for all users). Others accept preemption (preemption cleans up a resource for an exclusive usage) for users who have to deliver results at given times during the day. A good example is weather forecast that has to be ready at 6 pm such that it can be presented after the news at 8. This implies that the needed resources have to be reserved for the time needed to finish at 6, and this every day. All jobs running on those nodes at start time of the weather forecast must be checkpointed and restarted after 6. The CPU time of preempted jobs cost more, whereas the checkpointed jobs benefit from a cost reduction.

If priority can be used without preemption, it is necessary to define a very strict policy. In this case, a high priority job jumps ahead in the input queue, increasing the waiting time of all the jobs that are pushed back. As a consequence, higher priority should imply higher CPU costs, and lower CPU costs for all those jobs that end with higher turn-around times.

In the academic world (as at CSCS), a user often gets a certain monthly CPU time allocation. When this time is spent, the priority automatically is lowered. As a consequence, his jobs stay longer in the input queue, or, according to the local policy, he only enters a machine when the input queue of higher priority jobs is empty.

For the current implementation of ïanos, the priority is put to 1 for all C_k.

8.3.3 License fees K_ℓ

$$K_\ell(a_k, r_i, P_{k,j}) = \int_{t_{k,j}^s}^{t_{k,j}^e} k_\ell(a_k, r_i, P_{k,j}, t) \, dt \ [\text{ECU}]. \tag{8.11}$$

A license fee model is very complex. The simplest model is to directly connect the license fees to the CPU costs, $K_\ell = cK_e$. In some cases the computing center pays an annual fee and puts this fee into the CPU time, $c = 0$. Clearly, those users who do not use this program are not happy to pay for other users. Another simple model is to pay only if the program is really used. Then, the fee can directly be proportional to the CPU costs, $a > 0$. This model is applied when, for instance, the Computational Fluid Dynamics (CFD) code FLUENT is used in a project including academia and industry.

Note that the license often includes a maximum number of users (=tokens) who can simultaneously use a commercial code. If this number of reserved tokens is reached, a new user has to wait. This problem is first solved in the prologue phase. If there is no free token at t_0^k, then the machine is not eligible.

8.3.4 Costs due to waiting time K_w

$$K_w(t_k^0, t_{k,j}^e) = \int_{t_k^0}^{t_{k,j}^e} k_w(t) \, dt \ [\text{ECU}]. \tag{8.12}$$

This cost is user dependent. It could be the user's salary or a critical time-to-market product waiting cost.

Figure 8.4 shows an example of k_w concerning engineer's salary. Here, it is supposed that the user wastes his time only during working hours. A more sophisticated function could be yearly graphs also including unproductive periods like vacations. Figure 8.4 also shows an example of k_w of a critical time-to-market product.

But this cost has to be computed over all application components from the submission time to the end time of the last component running.

8.3.5 Energy costs K_{eco}

$$K_{eco}(a_k, r_i, P_{k,j}) = \int_{t_{k,j}^s}^{t_{k,j}^e} k_{eco}(a_k, r_i, P_{k,j}, t) \, dt \ [\text{ECU}]. \tag{8.13}$$

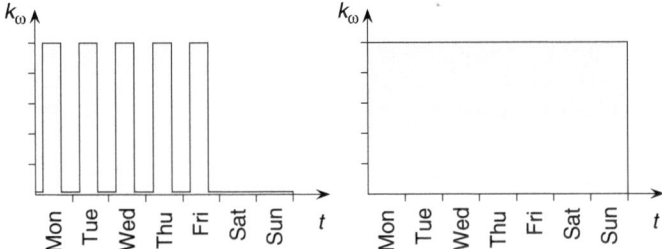

Fig. 8.4 Examples of waiting cost graphs. Left: Engineer's salary cost function $k_w(t)$ due to waiting on the result. Right: Time-to-market arguments can push up priority of the job

Energy costs over the lifetime of a node are a non-negligible part in the cost model. It enters strongly when the machine becomes old, and the investment costs become a small part of the CPU costs. For components that are memory bandwidth bound, the frequency of the processor could be lowered. First tests have been made with a laptop computer. When reducing frequency from 2.4 GHz to 1.2 GHz, the overall performance of a memory bandwidth bound application was only reduced by 2% (see 8.5). We have to mention here that for low-cost PCs energy costs (power supply + cooling) over 5 years can become comparable to investment costs. Thus, in future it is crucial to be able to under-clock the processor, adapting its frequency to the application component needs [88]. This could reduce the worldwide PC energy consumption, and free a few nuclear power plants. Computer manufacturers must be convinced to be able to have energy consumption graphs as the one depicted at the right of Figure 8.5.

The hourly energy costs for one node corresponds to

$$E_h^i = E_{node}^i + E_{cool}^i, \qquad (8.14)$$

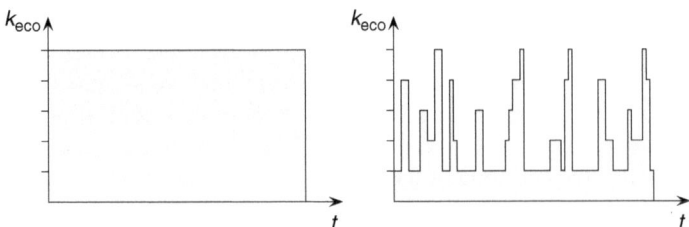

Fig. 8.5 Examples of graphs for the energy costs. Today (left): Excessive costs of energy consumption and cooling. Future (right): Energy consumption reduction due to frequency adaptation to application component needs

where E^i_{node} and E^i_{cool} are the hourly energy costs for the power supply and the cooling, respectively.

8.3.6 Data transfer costs K_d

Let us consider that different application components run on different servers located in different computing centers. The following data has then to be transferred between the different sites:

- Transfer of the component and its input data between the client and the computing center (client-server, cs)
- Data transfer between the different components (server-server, ss)
- Data transfer during execution to the client, for instance for remote rendering (server-visualization, sv)
- Transfer of the final result to the client (server-client, sc)
- Data transfers due to local or remote visualization.

Then:

$$K_d(a_k, r_i) = K_{d,cs}(a_k, r_i) + K_{d,ss}(a_k, r_i) + K_{d,sv}(a_k, r_i) + K_{d,sc}(a_k, r_i) \quad (8.15)$$

In Germany and Switzerland there is no precise model that estimates these K_d quantities. Presently, SWITCH (Swiss National Academic and Industrial Network) charges the traffic into the commodity Internet, but only during peak traffic periods (Monday to Friday, 08:00-20:00); 1 ECU/GB for academic users, 3 ECU/GB for others. In addition, there are flat rates for connecting to the Internet depending on the bandwidth (K_dc) (large universities have 10 Gb/sec) and size of the university (K_ds). In the case of a specific university that transfers about 160 TB/year, the mix of these costs results in an estimated GB transfer price of the order of 2.5 ECU/GB ($= 1\,\text{ECU} + (K_dc + K_ds)/160\text{TB}$).

8.3.7 Example: The Pleiades clusters CPU cost per hour

Let us give an example of how to determine the CPU and energy costs of the three Pleiades clusters. In Table 8.1 all the values representing costs are given in arbitrary units. A "k" after a number means "thousand". The interests S^i_b, the personnel costs S^i_p, and the management overhead S^i_a are distributed among the three machines according to the initial investment S^i_c. The infrastructure costs are distributed with respect to the number of nodes. For the Pleiades 1 machine y_1 has been chosen such that after 5 years the value of one node corresponds to the value of one Pleiades 2

Table 8.1 Characteristic parameters for the Pleiades clusters

Item	Pleiades 1 $i = 1$	Pleiades 2 $i = 2$	Pleiades 2+ $i = 3$
T_0^i	January 1, 2004	January 1, 2006	January 1, 2007
Nodes	Pentium 4	Xeon	Woodcrest
Architecture	32bits	64bits	64bits
OS	Linux SUSE 9.0	Linux SUSE 9.3	Linux SUSE 10.1
P_i	132	120	210
N_{CPU}	1	1	2
N_{core}	1	1	2
R_∞	5.6 GF/s	5.6 GF/s	21.33 GF/s
M_∞	0.8 Gw/s	0.8 Gw/s	2.67 Gw/s
V_M	7	7	8
Network	Fast Ethernet	GbE	GbE
y_i	1.5	2	2
r_i	1/year	1/year	1/year
T_i	4.5years	3years	3years
E_i	0.4 kW	0.4 kW	0.4 kW
u_i	0.8	0.72	0.76
F_i	0.1 /kWh	0.1 /kWh	0.1 /kWh
S_c^i	320k	270k	420k
S_m^i	20k	0	0
S_b^i	16k	14k	21k
S_p^i	100k	85k	135k
S_l^i	30k	28k	22k
S_a^i	50k	40k	70k
S_f^i	0	0	0
S_g^i	0	0	0
S_i	0.23	0.22	0.40
E_h^i	0.04	0.04	0.04
$\rho(i,1.1.2007)$	46k	93k	290k
K_ρ^i	0.05	0.12	0.47
K_i	0.32	0.38	0.91

node after 3 years. The idea behind is that a single node of Pleiades 1 and Pleiades 2 have the same performances, even though Pleiades 1 has been installed 2 years before. The quantity $\rho(i,\text{January 1, 2007})$ corresponds to the basis value of the machine i at January 1, 2007.

The result

$$K_i = K_\rho^i + S_i + E_h^i \qquad (8.16)$$

reflects the total hourly costs (investment, auxiliary, and energy) of one computational node at January 1, 2007, and K_ρ^i is the hourly node cost contribution due to the investment costs. The newest installation, Pleiades 2+, consists of the most recent Xeon 5150 *Woodcrest* nodes with two dual cores, where each node is 3 to 5 times more powerful than Pleiades 1 or Pleiades 2. This factor depends on the type of applications. Thus, from a user point of view, the Xeon 5150 *Woodcrest* machine is clearly the most interesting machine to choose, since four Pleiades 1 nodes or Pleiades 2 nodes cost about 50% more than one Pleiades 2+ node.

8.3.8 Different currencies in a Grid environment

The Cost Function Model is based on an arbitrary currency, the Electronic Cost Unit ([ECU]). We present in Table 8.1 an example of the values of the parameters that enter the model. These values are in Swiss Francs [CHF]. In fact, for a Grid with machines in different countries, Switzerland uses the [CHF] currency while Germany is in the Euro zone [128] where the currency is the Euro [EUR].

The scheduling decision is done at submission time (or some microseconds after). The models that drive the currency exchange done in an international market are not trivial but they must be entered into the cost model in the future.

8.4 The implementation

The ïanos implementation is a standard based, interoperable scheduling framework. It comprises four general web services and three modules:

1. the *Meta Scheduler* (MSS) performs resource discovery, candidate resource selection and job management
2. the *Resource Broker* is responsible for the selection of suitable resources based on a Cost Function model
3. the *System Information* is a front-end to Data Warehouse module, analyzes the stored execution data of a given application to compute certain free parameters used by scheduling models
4. the *Monitoring Service* passes the submission information received from MSS to the Monitoring Module and sends monitored data received from the Monitoring Module to the System Information
5. the *Monitoring Module* monitors the application during execution and computes execution relevant quantities
6. the *Data Warehouse* module is part of SI and stores information on applications, Grid resources and execution related data
7. the *Grid Adapter* module provides generic interfaces and components to interact with different Grid middlewares

The framework is complemented with one *ïanos client* that submits the application to the MSS using WS-Agreement, and the *Web Admin* that provides a web interface to store application and resource relevant parameters and data into the Data Warehouse.

ïanos allocates computing and network resources in a coordinated fashion [136]. The ïanos MSS prepares a list of candidate resources with their availabilities and sends it to the Broker. The Broker collects all the data needed to evaluate the cost function model, prepares a list of five potentially optimal schedules that is sent back to the MSS. If still available, the latter module submits the application to the machine with lowest costs. Otherwise, the second most cost efficient resource is chosen and so on. If all the five proposals are not available, the Broker is reactivated to recompute another set of proposals.

The implementation is based on state-of-the-art Grid and Web services technology as well as existing and emerging standards such as WSRF, WS-Agreement, JSDL, and GLUE. The ïanos is a general scheduling framework that is agnostic to a Grid middleware, and therefore can easily be adapted to any Grid middleware. The beta version of ïanos has been tested and integrated within UNICORE 5 by implementing a Grid Adapter for it. The present version includes all the models and monitoring capabilities. The free parameters in the cost function model have been validated on GbE clusters with well-known applications. The next step is to build a Grid of different type of resources and a fine tuning of the free parameters using a set of relevant applications coming from the HPCN community.

The ïanos middleware helps not only in optimal scheduling of applications but also to collect execution data on Grid resources. The monitored data on past application executions can be used to detect overloaded resources and to pin-point inefficient applications that could be further optimized.

8.4.1 Architecture & Design

The ïanos architecture is presented in Figure 8.2. In the following subsections the different modules are presented.

8.4.2 The Grid Adapter

The Grid Adapter mediates access to a Grid System through a generic set of modules as shown in Figure 8.6. It provides information on the Grid resources including CPU time availability and handles the submission of jobs on the selected resources. The SiteManager queries the Grid system for a list of available Grid Sites based on their type, for example, Globus site or UNICORE site. The InfoManager provides static information on the hardware and software configurations, on usage policies and dynamic information on resource availability, concerning the current loads. All this information is modeled as an *extended GLUE scheme* [8]. The TemplateManager contacts available Grid Sites for their installed applications. Each application is modeled as a *WSAG-Template*. The DataManager is responsible for the stage in/out of job files. The SubmissionManager is responsible for job submission and management. The ReservationManager handles the reservation of computational

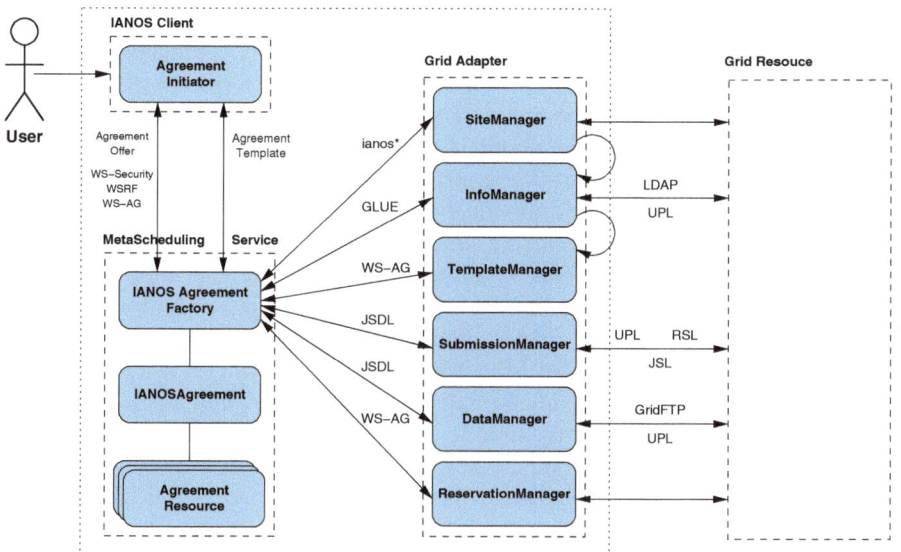

Fig. 8.6 Grid Adapter Modules & their Interaction with MSS and Grid Resource

and network resources for a job submission. The resources are reserved for certain duration of start and end times. As input, the JobManager and the DataManager receive a *JSDL* [9] while the ReservationManager receives *WS-Agreement*.

The Grid Adapter is the only ïanos module that is connected to a Grid system. Therefore, necessary interfaces are defined for each Grid adapter module to allow for easy integration with different Grid middlewares. We do not need to implement a new Grid Adapter for each Grid middleware. Instead, a plug-in mechanism is used in the design of Grid Adapter modules. At the moment, a Grid adapter plug-in for the UNICORE Grid middleware has been implemented.

8.4.3 The Meta Scheduling Service (MSS)

The MSS is the only ïanos service that is accessible by the client. It performs client authentication and candidate resources selection. It uses the Grid Adapter to access the Grid system. The access to the client is provided by an *AgreementFactory* interface. The client and the MSS communicate over the WS-Agreement protocol. The Client requests installed applications by issuing an *AgreementTemplate* request. The MSS first validates user requests and then queries the underlying Grid system for the list of installed applications based on user authorization filtering and application availability. The MSS sends these applications to the client in the form of *AgreementTemplates*. The client selects and submits the application to MSS by sending an *AgreementOffer*. This *AgreementOffer* includes a job description, application parameters and user QoS preferences such as maximum cost or maximum turnaround time.

Upon receiving an *AgreementOffer* from a client, it first identifies potential candidate resources and then queries the Grid Adapter for each candidate's resources static and dynamic information, which is received in *GLUE* format. It also retrieves the CPU time availability information (TimeSlots) for each candidate resource by contacting their local RMS. The submitted application is represented by an extended *JSDL* with information on intrinsic application parameters, and user QoS preferences. *GLUE* contains the information on candidate resources. The MSS sends JSDL and GLUE documents to the Resource Broker.

The response from the Broker is an ordered list of five execution configurations. Five is an empirical choice. Each configuration is represented by JSDL and includes start and end times of the execution, required number of nodes and the cost value for this configuration. The MSS starts negotiation with resources and uses the GA to schedule one of the configurations following the preferences expressed in the ordered list. If it is not possible to schedule any of the five configurations, the Broker is recontacted to compute new configurations based on the new availability information. A *WS-Agreement Resource* is created for each successfully submitted job on a selected resource. Job information is stored as resource properties. The MSS supports complete job management, such as job monitoring and control.

8.4.4 The Resource Broker

The Broker service carries out only one operation, *getSystemSelection*. The parameters of this operation are *JSDL* and *GLUE* documents which represent the candidate resources, the application parameters and the user QoS preference. The Broker uses two modules to decide on the suitable resources for a given application. The *EtemModule* implements the Execution Time Evaluation model and the *CFM Module* implements the Cost Function model. Applications can use the standard cost function implementation or a separate plug-in tailored to a specific application. This implementation framework allows separating implementations of different models and also to extend or provide new implementations of the models.

Each candidate's resources availability information is in the form of TimeSlots, where every TimeSlot is represented by the start and the end time, and by the free number of nodes during this available time. To compute a cost function value for each TimeSlot of the candidate's resources, the RB needs ïanos relevant resource parameters shown in Table 8.1, and data on the characteristics and requirements of the submitted application. The RB contacts the System Information for this data then based on the application requirements of nodes, memory or libraries, filters out unsuitable candidate resources. It then calculates the cost values for all suitable TimeSlots of the candidate resources (execution time is less than available time), prepares an ordered list of suitable configurations (including start-time and deadlines), and sends it to the MSS. Each configuration along with job requirements is represented by *JSDL*.

8.4.5 The System Information

The System Information exposes three operations: *getAppISSInfo*, *updateAppExecutionInfo* and *updateISSInfo*. The *getAppISSInfo* operation provides data on given resources and applications to the Broker. The *updateAppExecutionInfo* operation receives execution related data from the Monitoring service. The *updateISSInfo* operation receives ïanos relevant static resource information from Monitoring service.

The System Information is a front-end to the Data Warehouse module. An interface is designed to allow integration between System Information and Data Warehouse. This interface is independent of the specific implementation of the Data Warehouse. The Data Warehouse is contacted by the System Information to query, add or update stored data. It includes the ETEM Solver module that recomputes the execution time prediction parameters after each job execution. These parameters are then used by the Broker to predict the execution time for the next application submission.

8.4.6 The Data Warehouse

The Data WareHouse is a repository of all information related to the applications, to the resources found, to the previous execution data, and to the monitoring after each execution. Specifically, the Data Warehouse contains the following information:

- Resources: Application independent hardware quantities and ïanos related characteristic parameters
- Applications: Application characteristics and requirements such as software, libraries, memory, performance
- Execution: Execution data for each executed run which includes execution dependent hardware quantities, application intrinsic parameters

A Web Admin interface is designed to add or update information about resources and applications in the Data Warehouse.

8.4.7 The Monitoring Service

The Monitoring service receives submission information in JSDL from MSS and passes it to the Monitoring Module. This monitors the "ïanos" application. Upon receiving the monitored execution data from the Monitoring Module, it sends them to the System Information.

8.4.8 The Monitoring Module VAMOS

The Monitoring Module is called **V**eritable **A**pplication **MO**nitoring **S**ystem (VAMOS) [88]. During application execution, it measures and collects the application relevant execution quantities such as MF rate, memory needs, cache misses, communication and network relevant information. It maps these quantities which have been measured via the standard Ganglia monitoring system, with the informations from the local scheduler t_s and t_e. It is then possible to have an automatic profile of the application in a non-intrusive manner. Figure 8.7 shows an example of such a profile.

To have access to the MPI traffic a module has been developed. This module uses the PMPI API to count all the calls to the MPI API. The quantities are then sent to the Ganglia Round-Robin DataBase (RRDB) as a new metric and processed like the common Ganglia metric. VAMOS counts the size and number of in and out messages.

"VAMOS" and "full VAMOS"

To have access to low-level hardware events such as the number of *Flop* or the number of cache-misses, VAMOS integrates a module that uses the hardware counters available on the modern architectures. The counters are available on our IELNX cluster which runs the full VAMOS. The measured quantities (with the same time rate as the Ganglia monitoring system) is then sent to the Ganglia RRDB as a new metric. We use Perfmon2 to access the HW counters, PAPI is another usable API.

This low-level hardware events monitoring module can be switched on or off. If it is on, then we talk about **full VAMOS**, if it is off, the monitoring module is simply **VAMOS**.

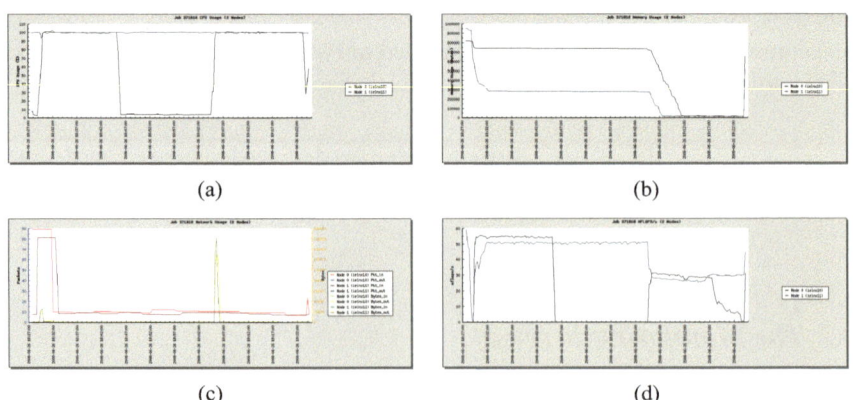

Fig. 8.7 A VAMOS example (SpecuLOOS running on 2 nodes of the IELNX cluster for 1 hour). (a) The CPU usage on the nodes from t_s to t_e. (b) The memory usage on the nodes from t_s to t_e. (c) The MPI traffic in and out from t_s to t_e. (d) The MFlops rate on the nodes from t_s to t_e

8.4.9 Integration with UNICORE Grid System

In a first phase of the ïanos project, we have integrated the ïanos framework with the UNICORE Grid System. The Grid Adapter with its generic functionality and the modules plug-ins specific for UNICORE, have been implemented and tested. On the client side, a UNICORE Client plug-in has been developed for this purpose. This client plug-in provides a GUI interface to interact with the MSS.

8.4.10 Scheduling algorithm

The scheduling algorithm implementation is in two parts: either it is a single node application or a multi-nodes application. Algorithm 4 is for a single node application while Algorithm 5 is for a multi-nodes application. Note that if the application is an embarrassingly parallel application without any internode communications, the multi-nodes algorithm is used.

Input: Single node application
Output: A schedule
estimate O for a ;
initialize a list of configurations $\mathcal{L}$;
foreach *Candidate Resource* r_i **do**
    get resource TimeSlots (availability info for r_i) ;
    estimate $T_{exec,predicted}(a, r_i)$;
    if *QoS Preference == Minimum_Time* **then**
        if $T_{exec,predicted}(a, r_i) \leq TimeSlot.getAvailableTime$ **then**
            compute $K_w(a, r_i)$;
            `/* ` $\alpha_k = \gamma_k = \delta_k = 0$ ` only ` $\beta \neq 0$`: the total cost is ` K_w ` */`
            compute $K = K_w$;
        end
    else if *QoS Preference == Minimum_Cost* **then**
        if $T_{exec,predicted}(a, r_i) \leq TimeSlot.getAvailableTime$ **then**
            compute K_e, K_l, K_d, K_{eco} ;
            `/* ` $\alpha_k, \gamma_k, \delta_k \neq 0$ ` and ` $\beta = 0$`: no need to compute ` K_w ` */`
            compute $K = \alpha(K_e + K_l) + \gamma K_{eco} + \delta K_d$;
        end
    else
        if $T_{exec,predicted}(a, r_i) \leq TimeSlot.getAvailableTime$ **then**
            compute $K_e, K_l, K_d, K_{eco}, K_w$;
            compute $K = \alpha(K_e + K_l) + \beta K_w + \gamma K_{eco} + \delta K_d$;
        end
    end
    add the configuration $conf = \{K, a, r_i\}$ to $\mathcal{L}$;
end
order $\mathcal{L}$ with respect to the cost from the lowest to the highest ;
return $\mathcal{L}$;

Algorithm 4: The scheduling algorithm for a single node ($NumNodes = 1$) application.

Input: Multi nodes application (parallel application) a
Output: A schedule
estimate O_k, S_k and Z_k for a_k ;
initialize a list of configurations $\mathcal{L}$;
foreach *Candidate Resource r_i* **do**
    get resource TimeSlots (availability info) ;
    if *QoS Preference == Minimum_Time* **then**
        foreach *TimeSlot of r_i* **do**
            P = TimeSlot.getAvailableNodes;
            /* Try all the possible values of P */
            while *$P > 0$* **do**
                estimate $T_{exec,predicted}(a, r_i, P)$;
                if $T_{exec,predicted}(a, r_i, P) \leq$ *TimeSlot.getAvailableTime* **then**
                    compute $K_w(a, r_i, P)$;
                    /* $\alpha_k = \gamma_k = \delta_k = 0$ only $\beta \neq 0$: the total cost is
                        K_w */
                    add the configuration $conf = \{K_w, a, r_i, P\}$ to $\mathcal{L}$;
                end
                $P = P - 1$;
            end
        end
    else if *QoS Preference == Minimum_Cost* **then**
        foreach *TimeSlot of r_i* **do**
            P = TimeSlot.getAvailableNodes;
            while *$P > 0$* **do**
                estimate $T_{exec,predicted}(a, r_i, P)$;
                if $T_{exec,predicted}(a, r_i, P) \leq$ *TimeSlot.getAvailableTime* **then**
                    compute K_e, K_l, K_d, K_{eco} ;
                    /* $\alpha_k, \gamma_k, \delta_k \neq 0$ and $\beta = 0$: no need to compute K_w
                    */
                    compute $K = \alpha(K_e + K_l) + \gamma K_{eco} + \delta K_d$;
                    add the configuration $conf = \{K, a, r_i, P\}$ to $\mathcal{L}$;
                end
                $P = P - 1$;
            end
        end
    else
        foreach *TimeSlot of r_i* **do**
            P = TimeSlot.getAvailableNodes;
            while *$P > 0$* **do**
                estimate $T_{exec,predicted}(a, r_i, P)$;
                if $T_{exec,predicted}(a, r_i, P) \leq$ *TimeSlot.getAvailableTime* **then**
                  compute $K_e, K_l, K_d, K_{eco}, K_w$;
                  compute $K = \alpha(K_e + K_l) + \beta K_w + \gamma K_{eco} + \delta K_d$;
                  add the configuration $conf = \{K, a, r_i, P\}$ to $\mathcal{L}$;
                end
                $P = P - 1$;
            end
        end
    end
end
order $\mathcal{L}$ with respect to the cost from the lowest to the highest ;
return $\mathcal{L}$;

Algorithm 5: The scheduling algorithm for a multi nodes (*NumNodes* > 1) application.

8.4.11 User Interfaces to the ïanos framework

Two *Graphical User Interfaces* are offered to operate the ïanos framework implementation as two actors exist: the user and the system administrator on the computing centers side.

The user interface

In order to submit a job through ïanos, a user u_i must install a thin client implemented as a UNICORE client plugin. Figure 8.8 shows an example of such a client.
The current version of the client is as simple as possible. The user first *gets the
available applications* (an available application is an application the user is allowed
to run). Then he chooses an application. He gives the input parameters through the
GUI or by files (the size of the matrices for the DGEMM application, for instance).
Finally he must give the QoS. Four choices are available and implemented: (1) *Min-
imal Cost*, (2) *Minimum Time*, (3) *Optimal* (trade-off between the two first) and (4)
the *expert* mode where the user can fine tune each weight α, β, γ and δ of the cost
function. When all is selected, the user is allowed to submit the application. During
execution, the user can monitor his job in the sense of timing. He is also allowed to
kill the job.

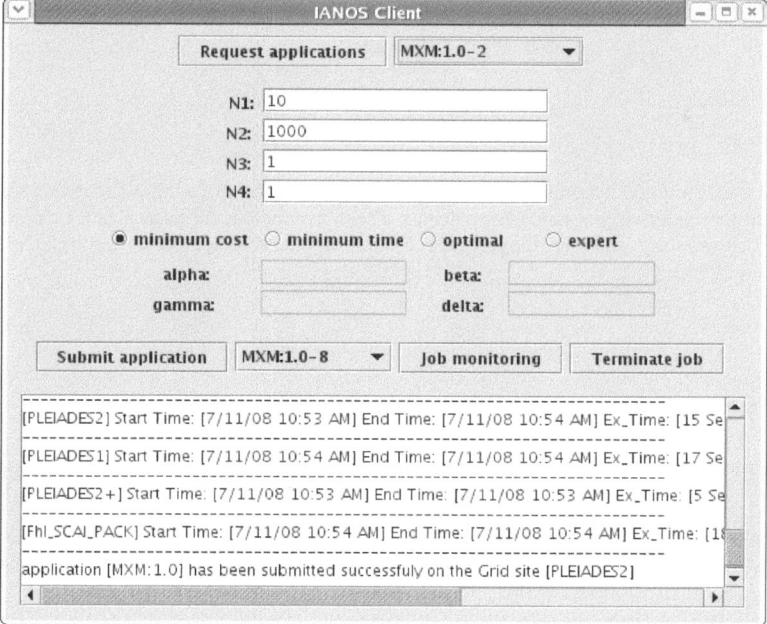

Fig. 8.8 The ïanos client. It authorizes the user u to choose among the application he is allowed to
run, to give the related parameters (N_i) and the Quality of Service he wants (values of α, β, γ and
δ or pre-selected values). The sub window shows the scheduling process and the chosen resources

The web interface for ïanos's administration

In Chapter 3 we define a Grid site (see Definition 2) as *a set of resources administered and managed according to local policies by one administrator (or administration). Each site can have its special policy of usage*. Thus, all the site related
quantities must be fulfilled by the local administrator. These quantities are those
described in Table 8.1. The implementation choice is a Web User Interface that
gathers all the information. Figure 8.9 is a snapshot of the web Admin interface for
the Pleiades clusters.

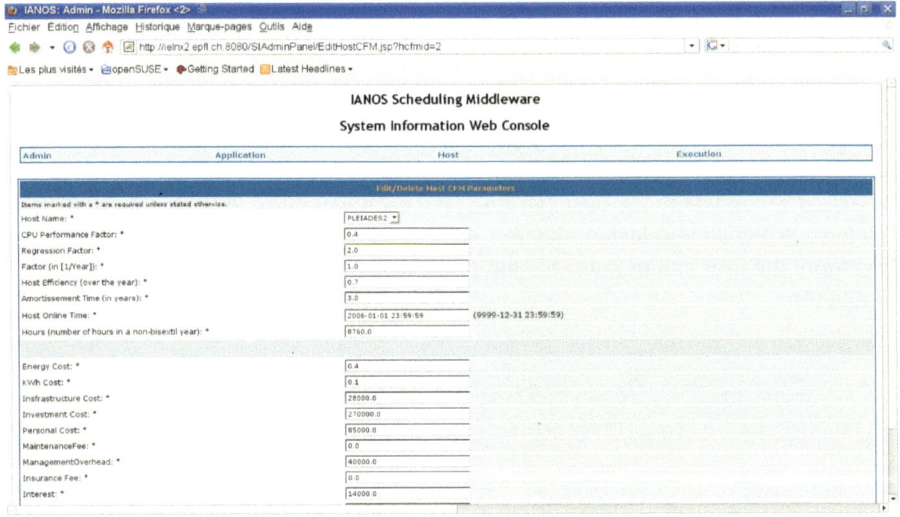

Fig. 8.9 The ïanos web administration interface. The administrator of a ïanos site must give information on the underlying systems he manages. These parameters are used in the Cost Function
Model to estimate the Cost, thus the schedule, of jobs onto the resources. A resource can be added,
edited or removed

8.5   DVS-able processors

Robert Blackburn proposes five ways to reduce energy consumption in the data centers [17]: (1) identify the culprits, (2) enable server processor power saving features,
(3) right-size server farms, (4) power down servers when not in use, and (5) decom-
mission old systems that provide no useful work. In this appendix, we'll concentrate
on the second point.

Mark Weiser *et al.*   [137] introduce an energy-aware metric for processors:
the *millions of instructions per Joule* (MIPJ) and propose original methods to
reduce MIPJ on desktop computers. With the appearance of mobile computing, this
topic became important for battery saving. "Energy Profilers" have been developed

to pinpoint the energy-eager processes in a system [51] used to better schedule applications on a mobile platform. Bianchini and Rajamony [16] show that 22% of the energy consumption of a single server is needed to cool it.

Today, it is common to talk about *power* rather than *energy* (power = energy per unit of time). The Top500 list introduced recently (June 2008) the total power consumption of the supercomputer facilities in the list. The Green500 introduces the *Flops/Watt* metric [115].

The model we present here assumes that the dominant part of the power consumption of a resource is the CPU. We are aware that the other *static* parts of the resource also consume energy, and sometimes more that the CPU. David C. Snowdon *et al.* show the impact of the memory consumption in [117], Seung Woo Son *et al.* in [118] present a model to reduce energy for sparse matrix based applications running on a distributed memory based system.

8.5.1 Power consumption of a CPU

IT literature gives several models to calculate the CPU power consumption. If we assume that a CPU is a CMOS, the power consumption of the chip is given by Eq. (8.17) [107] .

$$P = \underbrace{ACV^2f}_{P_1} + \underbrace{\tau AVI_{short}}_{P_2} + \underbrace{VI_{leak}}_{P_3}, \tag{8.17}$$

where A is the activity in the gates in the system, C the total capacitance seen by the gate outputs, V the supply voltage, f the operating frequency. The quantity P_1 measures the dynamic power consumption, P_2 measures the power spent due to short-circuit currents I_{short}, P_3 measures the power lost due to leakage current (a small undesirable flow of current through the chip) that is present regardless of the state of the gates. The quantity P_1 is dominant, Eq. (8.17). Thus,

$$P \approx ACV^2f \tag{8.18}$$
$$\approx V^2f .$$

The current processors still behave as shown in Eq. (8.18). It is not the purpose of this book to go into details of the architecture of the chips but a small survey follows. A core c of a CPU has a clock with a core frequency f_0. For the current CPU's, f_0 corresponds to 50 MHz to 250 MHz. A special device – the multiplier – increases or decreases the CPU core frequency f_0 by a factor of r_f: $r_f \times f_0$. The multiplier increases and decreases also the processor core voltage V_{cc} by a factor r_v: $r_v \times V_{cc}$. The quantities r_f and r_v are called **bus-to-core ratios**. We then talk about **P-State** for the couple $(r_f; r_v)$ and **C-State** for a couple $(r_{f,c}; r_{v,c})$ related to a core c: modern processors can have different clock rates and supply voltages per core. The exact values of the P-States and C-States are not publicly available.

The chip founders implemented a technology to change the values of P-States (or C-State) dynamically. This power management technique is called Dynamic Voltage Scaling (DVS) [81, 4]. The technology allows to influence the frequency, not the voltage. The latter is determined by the P-States (C-States). The DVS implementation is called *SpeedStep*© by Intel and *Cool'n'Quiet*© by AMD.

8.5.2 An algorithm to save energy

In Eq. (3.10) (Chapter 3.3), g_M (defined as $M_m = g_m M_\infty$) is the relation between real to peak main memory bandwidth. If $V_a < \frac{V_m}{g_m}$, Eq. (3.18), the application is memory accesses dominated and does not need the full speed of the processor. We show in Section 4.5.3 that the SMXV is of this type. Furthermore, we describe in Section 8.4.8 an implementation of a Monitoring Module called VAMOS able to monitor R_a (thus V_a) in an automatic manner.

The *optimal operating frequency* f_a for such an application a is given by Eq. (8.19).

$$f_a = M_\infty * V_a \tag{8.19}$$

If the application a runs on the core c clocked at frequency f_a, almost no performance will be lost but energy could be saved. The resource r_i is tuned to the needs of the application a.

For the algorithm (Alg. 6) it is assumed that the application previously ran on a resource and that the information is stored in the SI (see Section 8.4.5). If the application is main memory access dominated (thus if $V_a < \frac{V_m}{g_m}$), we can compute the frequency of the core by Eq. (8.19).

Input: an application a and a resource M
Output: the *optimal operating frequency* f_a for a on the resource M

$g_M \leftarrow \texttt{getGMFromSI(a)};$
$V_a \leftarrow \texttt{getVaFromSI(a)};$
$V_M \leftarrow \texttt{getVMFromSI(M)};$
if $(V_a < \frac{V_M}{g_M})$ **then**
   | $M_\infty \leftarrow \texttt{getMinfinityFromSI(a)};$
   | $f_a = M_\infty * V_a$;
else
   | $f_a = f_\infty;$
end
return f_a

Algorithm 6: Frequency scaling based on the g_M quantity. Here, we assume that the application a already ran on the resource M, thus the application's related quantities g_M and V_a are available in a System Information. The resource's quantities V_M and M_∞ are also stored in the System Information

In Section 8.3.5, we present the energy costs K_{eco} part in the cost function formulation. As far as we know, no resource brokering algorithm takes into account the energy cost; thus no model exist. Using the *optimal operating frequency* f_a for an application on a resource r_i it is possible to improve the usage of energy-aware resources and lead the choice of the broker to resources that implements a DVS technology.

8.5.3 First results with SMXV

The memory access dominated SMXV application with $V_a = 1$ is chosen as a show-case (Section 4.5.3). We use a LENOVO T61 laptop with an Intel Core2Duo T7700 processor for the tests. The *cpufrequtils* package from the Linux Kernel repository is chosen to tune the frequency [27]. We estimate the voltage on the basis of the specifications[2] (1.0375 - 1.30V) and assume that the stepping is linear with frequency. Table 8.2 shows the estimation of the frequency, the voltage and the power consumption of the processor.

The SMXV application has been executed at all the frequencies in Table 8.2, the performances are measured, and the power consumption estimated. The results are presented in Table 8.3. Note that the Linux Kernel runs at *runlevel* 2 (without the XWindow system). The network was shut down. We note that the *optimal operating frequency* f_a given by the algorithm Alg. 6 is

$$f_a = (0.800 \cdot 1) = 0.800 \; GHz \, . \tag{8.20}$$

At this frequency a performance loss of 30 % is measured, more than expected. In practice, the real frequency is predicted by another algorithm (patent pending) in which the acceptable performance loss can be imposed.

Table 8.2 Estimation the voltage and the power based on frequency stepping by 400 MHz starting at the lowest sleeping frequency at 800 MHz, and the Intel T7700 specifications. The specifications specify that the peak power consumption of the processor is 35 $Watt$ thus $A\,C = 8.63$, Eq. (8.18)

core frequency GHz	core voltage V	Power W
0.8000	1.0375	7.45
1.2000	1.1031	12.63
1.6000	1.1687	18.91
2.0000	1.2343	26.23
2.4000	1.2999	35.00

[2] http://processorfinder.intel.com/details.aspx?sSpec=SLA43

Table 8.3 Results of the SMXV application's building block on a LENOVO T61 equipped with an Intel T7700 processor operating at different CPU frequencies. For this example, $N_1 = 1$, $N_2 = 2000$. The energy is calculated using the power computed in Table 8.2 and the measured execution time T_{exec}

core frequency GHz	T_{exec} s	R_a GF	$\frac{R_a}{R_{a,max}}$	Energy J
0.8000	6.74	0.2671	0.66	50
1.2000	4.53	0.3974	0.98	57
1.6000	4.50	0.4000	0.99	85
2.0000	4.46	0.4036	0.998	117
2.4000	4.45	0.4045	1	156

8.5.4 A first implementation

Measurements of a first implementation on the LENOVO laptop are shown in Fig. 8.10 where the frequency, the temperature, and the processor performance are periodically measured during one hour clock time. The algorithm was on during the first 40 minutes, then it was switched off, and the normal scheduling was running. With the algorithm on, the laptop ran productively at 800 MHz during the periods

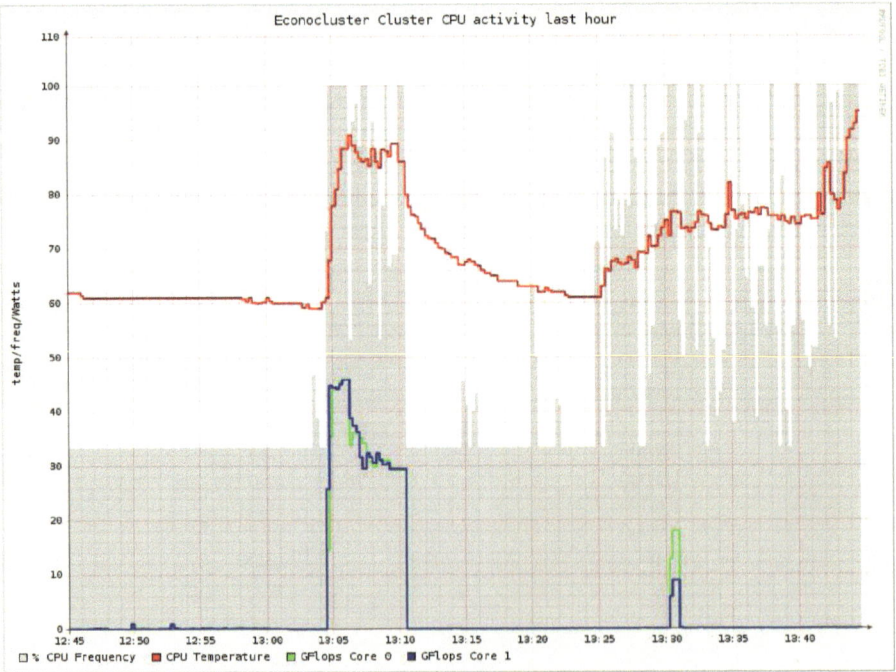

Fig. 8.10 First frequency tests on a LENOVO laptop. The running applications are random, the CPU frequency governor is "ondemand". The solid red line is the temperature, gray area is the frequency and the blue and green lines are the core performance for the two cores

when the user was working with text editors, emailing, and internet searches. When executing the DGEMM matrix times matrix operation, the algorithm forces the processor to run at top 2.4 GHz speed until the end of the execution. Afterwards, 800 MHz are again sufficient. However, if the energy saving software is switched off, the processor runs at full speed without any reason.

During the DGEMM execution a surprising frequency oscillation of the processor has been detected. Even though the 2.4 GHz were imposed by the algorithm, the processor changed automatically its frequency. The result was a unexpected performance reduction of the running application. The user found out that the processor cooling was reduced by accumulated dust. After cleaning the laptop, this performance reduction disappeared (Figs. 8.11 and 8.12).

Figs. 8.11 and 8.12 show the frequency evolution without and with the algorithm when running first the DGEMM kernel followed by the SMXV kernel. During the rest of the time no activity was requested. One sees that without the algorithm the processor ran at top frequency with a temperature of 80° during the execution of both kernels, whereas with the algorithm the frequency was reduced during execution of the SMXV program with a temperature of the processor at 50°. Not only

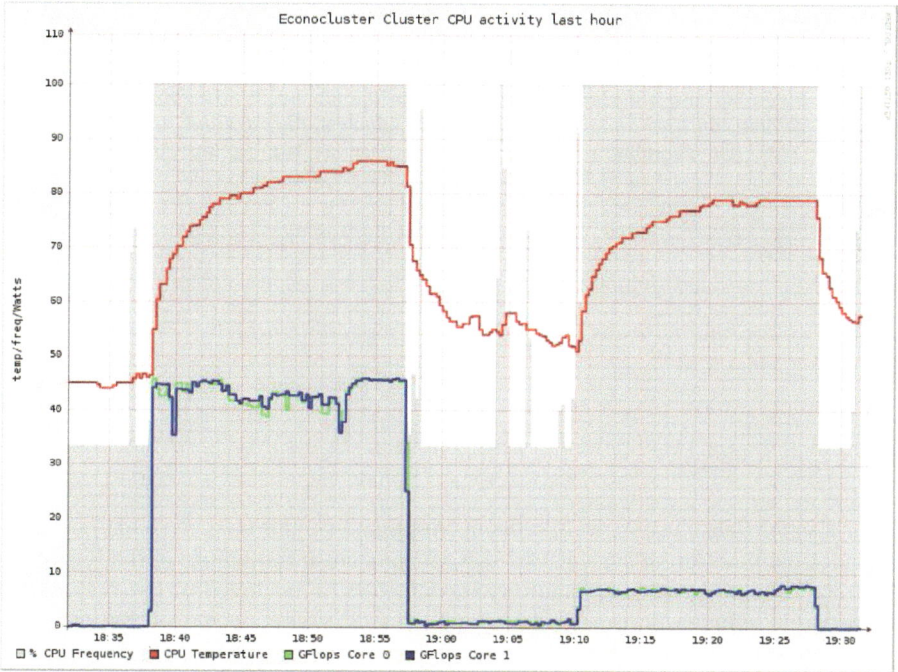

Fig. 8.11 First frequency tests: the frequency reduction algorithm is off. The computing intensive DGEMM computation lasts from 18:38 to 18:58 (20 minutes), the processor should then run at top. The memory bandwidth dominated kernel SMXV lasts from 19:10 to 19:28 (18 minutes), the processor should not run at top frequency, but it does ("conservative" frequency manager from the Linux kernel). The temperature (red plot) grows exponentially when the frequency is maximal

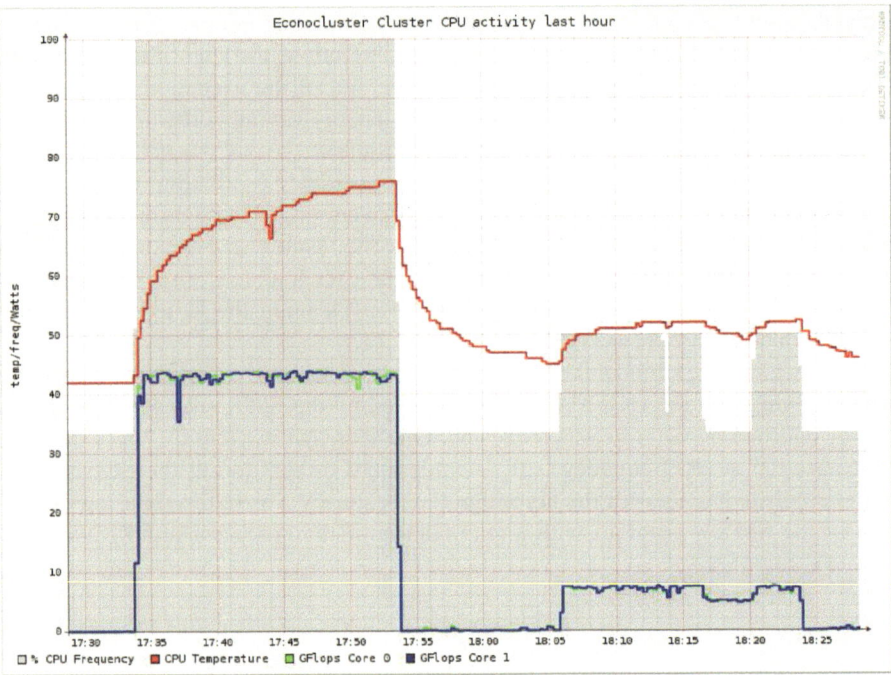

Fig. 8.12 First frequency tests: the frequency reduction algorithm is on. The computing intensive DGEMM computation lasts from 17:33 to 17:53 (20 minutes), the processor should then run at top. After 18:00, the algorithm reduces the frequency again to its minimum, 800 MHz, and the temperature drops exponentially back to 60°. During the memory bandwidth dominated kernel SMXV (between 18:05 and 18:23 thus 18 minutes), the frequency algorithm selects an appropriate frequency. The temperature (red plot) stays at 60°. Energy is saved without a lost of performance (see Figure 8.10)

the energy consumption was smaller but the temperature drop by 30° increases the MTBF by a factor of 8.

8.6 Conclusions

This chapter introduces the Intelligent ApplicatioN Oriented System, ïanos. It includes the monitoring system VAMOS, the System Information SI, the Resource Broker RB, and the MetaScheduling Service MSS. The choice of the best suited resources at submission time is done through a cost function model that includes costs related to CPU, licensing, data transfer, waiting time, and ecological arguments according to a QoS given by the user. The QoS is for instance *minimum turn around time*, *lowest costs*, or *minimum energy consumption*.

The monitored information can be used to adapt the processor frequency to application needs, to pinpoint machines that should be decommissioned, and to help managers to choose new resources for a given application community.

Chapter 9
Recommendations

"Advice is what we ask for when we already know the answer but wish we didn't."

Erica Jong, American writer

Abstract Here, we compile a few advices on how to design and write a code, and get it run fast. A special emphasis is made on proposing solutions to reduce energy consumption by adapting a computer to application needs, by choosing a well suited machine to run each application component, and by installing a Grid of resources that is best suited to the application community. We know that these advices could well be wrong.

9.1 Application oriented recommendations

9.1.1 Code development

In code development fundamental choices are necessary to produce results as precise as possible in a given time. Modular programming, a clear data organization, and a nice user interface help to let a code survive for a certain time, and encourage its usage. A program operates on data. Coding should be as simple as possible, no tricks, just standard programming. Then, the program is more easily be portable on another machine, efficiently be parallelized, the compiler is able to well optimize, and results are produced in a minimum of time.

9.1.2 Code validation

Code validation is very critical, but not discussed in the book. Here an example: Three research teams work on the same brand new subject and develop a new simulation code. They decide to compare the three codes by computing the same test case on which they agreed. Two teams get results that are close, the third produces different results. Who is right? The third team decided to apply a second, mathematically different, method to compute the same case, and gets twice the same answers. Later on, the other two teams corrected their codes.

A similar problem is the interpretation of results and their understanding. For this purpose, it is needed to well know the underlying physical and numerical models,

R. Gruber, V. Keller, *HPC@Green IT*,
DOI 10.1007/978-3-642-01789-6_9, © Springer-Verlag Berlin Heidelberg 2010

and the algorithms that have been chosen to describe them. Did the computation converge? Did I capture the right solution? Graphical representations of results are very useful, but can sometimes lead to misleading interpretations.

9.1.3 Porting codes

There is a large number of simulation programs that are outdated. They have been written and optimized for vector machines, and often use non-standard specificities introduced in old computer architectures to minimize main memory usage. When porting these codes on modern computers, the compiler sometimes refuses to compile, or what is even worse, wrong results come out. The advice is to redevelop new codes that follow well established standards, and that deliver parallelized and optimized executables.

9.1.4 Optimizing parallelized applications

The computer architectures become more and more complex. Large numbers of cores form a processor, many processors give a node, and clusters can include a huge number of nodes interconnected by efficient internode communication networks. In addition, one talks about Grids of parallel machines, each machine having its own resource characteristics. Such complex hardware environments makes it possible to optimize application execution. By cutting an application in modular components with small intercommunication needs, it is possible to choose for each component a well suited hardware platform for execution.

Each component can individually be parallelized on a well suited cluster using a standard internode communication library. To minimize latency time, the number of messages should be minimized, and data has to be assembled before the transfer.

The many core node architecture can be optimally used through OpenMP directives that are efficient if the compiler is able to keep the high level of core optimization, but this is not always possible. In fact, OpenMP adds instructions in the code to build parallel threads. This can complicate the code so much that the compiler is not further capable to well optimize for a single core. Then, the user can help the compiler by simplifying the coding, by breaking up too complicated loops, by reconsidering OpenMP parallelization, or by replacing OpenMP by MPI.

9.1.5 Race condition

When parallelizing applications, one danger is the race condition: An error can occur that is difficult to detect. Data can be read or written too early or too late, the simulation continues, and the result is not quite right. If the same case is run a second time, the results can be differently wrong. Since a user has the tendency to interpret a wrong result as correct; he would not run the same case a second time and miss the race.

Wrong results can be produced in different levels of parallelization. Components can read boundary conditions too early, others can write them too late. We have shown in chapter 6 a very simple race condition occurred with OpenMP that was easy to detect. In more complex circumstances, it can be difficult to recognize a race condition. If you want to be more confident, it is perhaps better to use MPI with enough barriers. Since the user has to organize the local data flow, race conditions only occur if he programs them explicitly. The best advice we can give here is to consult parallel application specialists.

9.2 Hardware and basic software aspects

Grid aware systems like Globus or UNICORE exist that are able to submit application components to the desired resource in a Grid. Very recently, the application component aware middleware ïanos is proposed to find resources in an automated manner that are well suited for each individual component.

Parallelization of distributed memory architectures can be realized by using the message passing communication library MPI. All the data transfers between nodes is done by passing messages. Parallelizing codes is an art. To reduce errors, it is advised to employ parallel programming specialists in a software development team, and let them do this critical work.

9.2.1 Basic software

We advise the readers to use existing kernels and library routines. They are well written, and optimized for all HPC resources. Examples of such basic softwares are the BLAS2 and BLAS3 kernels, or mathematical libraries such as ScaLAPACK (www.netlib.org), PETSc (www.petsc.org), FFTW (www.fftw.org), or MUMPS (www.mumps.org). It is also advised to use standard open ware data base systems like HDF5 (www.hdf5.org). Since those pieces of fine software are available on most compute platforms, applications including standard software packages are easily portable, and immediately run fast.

The internode communication standard is presently the MPI message passing library. In a node, the multi processor, many cores architecture can be used through OpenMP directives, but MPI is also possible. It is a sort of a dream that the old autotasking concept could be reintroduced in the compilers such that parallelism in a node will be automatically done in an optimal manner. The new CUDA programming standard has been developed for the graphics processors that slowly penetrate the HPC market.

9.2.2 Choice of system software

Again, it is advised to use a standard system like Linux. If the system is installed and works, then it is advised to not touch it again. Other basic softwares that should be installed in a standard manner are one to distribute a software prepared on one node to all the others, or to schedule the machine.

The choice of the compilers is more critical. The open software Gnu compilers (www.gnu.org) are standards, and available on all important computers. The new Intel compilers have recently been improved so much that it is advised to use them on the Intel machines. These compilers are now aware of the special SSE hardware, and can make use of it. Improvements by of up to a factor of 3 could be measured for processor dominant applications.

9.3 Energy reduction

Energy consumption due to IT constantly goes up and should be stopped. This can be done by adapting the hardware parameters to application needs, by improving the cooling mechanisms, by optimizing jobs, by choosing most adequate computing resources, or by replacing old resources.

9.3.1 Processor frequency adaptation

By periodically monitoring the behavior of the applications, it is possible to tune the processor frequency. The energy consumption is approximately proportional to the frequency and to the square of the applied voltage. Processor frequency adaptation is efficient for applications that are dominated by main memory access, when searching in data bases or over the Web, when reading from and writing on disks, or when waiting for data input if one writes a letter, reads and writes email, if one googles, or if an application waits for data input.

To be able to control a node, an application monitoring system has to run on the resource without influencing the run time of the application. According to the measured data, decisions on the processor frequency are taken periodically. The time period can be chosen, is typically one or several seconds. Well controlled frequency and voltage reductions give rise to application performance reduction of 2% or less, whereas energy reduction of the processor is measured go up to 50%, and of 10% to 20% for the whole node. This is clearly economically advantageous since energy in 5 years accounts at similar expenses as the investment costs. Thus, it is better to add 2% more computers, and reduce energy consumption without reducing overall performance of the resources. For such applications over-clocking should be stopped.

Not all the applications can profit from a reduction of energy consumption through frequency adaptation. The best known example is the Linpack benchmark that reaches typically 70% of peak processor performance. This is also true when using the special (SSE) hardware added in some computers to improve CPU time consuming matrix operations.

With the upcoming change in the ecological awareness, it could well be that application performance reduction will be accepted when the processor frequency and the voltage are reduced.

The energy consumption is not only coming from the processors. Processors consume perhaps one third of the energy, but the other parts, main memory, motherboard, graphics card, I/O subsystem, or the network communication also consume, altogether two thirds. In future, it is important to control all parts of a resource to optimally adapt a resource during application execution.

9.3.2 Improved cooling

Three cooling mechanisms are known: Air cooling, liquid cooling, and two phase cooling. The efficiency of these cooling mechanisms is related to the heat transfer. Heat conduction is 20 times larger in water than in air, its efficiency is much better, and energy for cooling can be saved. In addition, the temperature of the chips can be reduced and the MTBF increased. The heated liquid is cooled down in a heat exchanger, and the energy can be reused to warm up water or heat buildings. Since liquid cooling is expensive, it is only done in expensive servers; but they can then be integrated more densely.

Even better efficiencies could be reached when a phase transition can be used to cool an electronic chip. On the hot chip surface a two phase liquid absorbs the heat by passing from the liquid to the gaseous state at relatively low temperature around 50°. The gas is transported to a heat exchanger where it is liquefied. The exchanged energy can again be used to warm up water and/or buildings. The additional energy needed in this process is smaller than for liquid cooling, and very much smaller than for air cooling. The MTBF would go further up and the number of crashes down.

Generally spoken, energy recovery for producing warm water and for heating buildings would be highly efficient in cold regions where heating is needed during most of the time. Computing centers should be built in cold regions, liquid cooling applied, and energy reused.

9.3.3 Choice of optimal resources

If there are different types of resources available to run an application, it is advantageous to try to reserve the best suited machine at a given time. Best suited here means that the costs are as small as possible. These costs can include the CPU time costs, the costs due to communication, licensing of a commercial software, energy costs, or the waiting time that can reflect engineer's salary, or a later appearance of the product on the market.

9.3.4 Best choice of new computer

A computing centre can also use the historical data to get a hint on the resources that should be decommissioned, and on new resources that should be purchased for its application community. In one case, an older resource has been detected for which the energy consumption costs were larger than the overall costs when running on a modern machine. This older machine has shortly afterwards been replaced.

When purchasing a new computer, computer vendors have the tendency to offer the best Linpack machine to get high up in the TOP500 list. Thus, as many cores as possible are often proposed. We even experienced that machines with less cores had higher prices. Energy consumption is now becoming an argument in the choice of a resource. The reason is clear: If the target applications do not need these additional cores, having less cores diminishes energy costs, and, as a consequence, computing costs. For this purpose, it is necessary to better understand how to adapt computer architectures to application needs. Instead of a monolithic massively parallel machine highly ranked in the TOP500 list, it could be more profitable to have two or more complementary computer architectures that are better covering the needs, produce more results faster, and consume less energy.

9.3.5 Last but not least

Some computer manufacturers have understood that main memory bandwidth is very important for a big number of applications. From 2003 to 2009 they increased the peak main memory bandwidth by a factor of 10. However, in one case, the efficiency decreased from (already low) 50% to 18% when running on a single core and to 25% when running on all the cores. The main memory subsystem consumes most energy; a special radiator is mounted on top of it to cool it down. If one would run at half frequency and increase efficiency to 50% (still low), the energy consumption would fall by a factor without changing the bandwidth.

9.4 Miscellaneous

9.4.1 Course material

One of the authors has given a High Performance Computing Methods course at the EPFL during many years. Most of the collected material has been integrated in this book. The course is available upon demand.

9.4.2 A new REAL500 List

Chip founders such as Intel or AMD know that the market is driven by the Microsoft Windows operating system and its inefficient backward compatibility. But they also know that high performance computing (which represent around 30% of their sales) is viewed as a showcase. HPC is to computing what Formula One is to general public cars. If the engine of the number One of the TOP500 List (or GREEN500 List) is an Intel or an AMD, it is obvious that the company will use it intensively for marketing.

We showed in previous chapters that the TOP500 focuses only on BLAS3-dominated applications, thus is a direct measure of the peak processor performance. We propose to extend the TOP500 towards a REAL500 list that includes:

1. HPL benchmark as in the present TOP500 list
2. A benchmark that measures the real main memory bandwidth by a benchmark such as SMXV (Sparse Matrix*Vector multiply)
3. The total energy consumption including cooling. This GF/J rate should be given for the HPL and the SMXV benchmarks.

Glossary

A	Speedup
a	Application
$\mathcal{A}$	All applications in a Grid
a_k	Application component
a_{sd}	Task (or subdomain) of an application component
ABC	Atanasoff-Berry Computer
ALU	Arithmetic Logic Unit
API	Application Programming Interface
am_{core}	Addressing mode (32 or 64bits)
APL	First parallel compiler
availability	Holes in the input queues of r_i
b_i	Effective per node internode communication bandwidth
B_i	Communication bandwidth taking L_i for transfer
Beowulf	Cluster concept based on commodity components
BLAS1	Vector times vector Basic Linear Algebra Subprograms
BLAS2	Matrix times vector Basic Linear Algebra Subprograms
BLAS3	Matrix times matrix Basic Linear Algebra Subprograms
Blue Gene/L	MPP computer by IBM running at 700MHz
Blue Gene/P	MPP computer by IBM running at 800MHz
C_i	Total peak network bandwidth
C_∞	Peak network bandwidth of one link
CD	Compact disk
CDC 6600	Mainframe computer by CDC
CELL	PS3 node by IBM
CFD	Computational Fluid Mechanics
CFM	Cost Function Model
CM-5	Connection machine by TMC
configuration	Submission pattern for a_k on r_j
CPU	Central Processing Unit = Processor

R. Gruber, V. Keller, *HPC@Green IT*,
DOI 10.1007/978-3-642-01789-6, © Springer-Verlag Berlin Heidelberg 2010

Cray-1	First vector computer by Cray Inc.
Cray-2	4 processor vector computer by Cray Inc.
Cray X-MP	4 processor vector computer by Cray Inc.
Cray T3D	MPP computer by Cray Inc.
Cray T3E	MPP computer by Cray Inc.
Cray XT3	MPP computer by Cray Inc.
Cray Y-MP	8 processor SMP machine by Cray Inc.
CRT	Cathode Ray Tube

d	Number of days per year
$\langle d \rangle_i$	Average distance between *nodes*
D_i	Distance: Minimal longest path in a network
DARPA	Defense Advanced Research Project Agency
DDR	Dual Data Rate
DDR2	Double DDR
DDR3	Quadruple DDR
DDRAM	Dual Dynamic Random Access Memory
Deep Blue	Chess specific supercomputer by IBM
DGEMM	BLAS3 full matrix times matrix operation
DGEMV	BLAS2 full matrix times vector operation
DIMM	Dual Inline Memory Module
DMA	Direct Memory Access
DSP	Digital Signal Processor
DVS	Dynamic Voltage Scaling

E^i_{cool}	Energy consumption for cooling for resource r_i
E	Efficiency A/P
E^i_{node}	Energy consumption of node in resource r_i
ECU	Electronic Cost Unit
EDSAC	Electronic Delay Storage Automatic Calculator
EDVAC	Electronic Discrete Variable Arithmetic Computer
EIB	Element Interconnect Bus in a CELL
EM	Electromagnetic
ENIAC	Electronic Numerical Integrator and Computer
ERA	Computer company, later Univac
ERMETH	Early computer built at ETHZ
ES	Earth Simulator by NEC
ETEM	Execution Time Evaluation Model

F_{core}	Frequency of a core
f_0	processor core frequency
Ferranti	Early British computer company
FFT	Fast Fourier Transform
Flop (F)	Floating point operation

Flowmatic	First compiler
FPGA	Field-Programmable Gate Array
Γ	Node performance over network bandwidth
γ_a	O/S
γ_m	r_a/b
g_m	Efficiency of the main memory subsystem
g_p	Efficiency of the processor
GB	Gigabyte = 10^9 Bytes
GbE	Gigabit Ethernet
gcd	greatest common denominator
GF/s	10^9 floating point operations per second
GHz	10^9 Hertz (cycle periods) per second
Globus	Grid middleware
GLUE	Grid Laboratory for a Uniform Environment
Grid	A set of remote located sites
Grid Adapter	Interface to a site

H_i	Bisectional bandwidth
Hapertown	Intel Xeon node with $N_{CPU}=2$, $N_{core}=4$
HDD	Hard disk drive
HPC	High Performance Computing
HPL	High Performance Linpack benchmark for TOP500 list

i	Counter for resource
IC	Integrated Circuitry
I/O	Input/output system connected to disks or archives
ïanos	Intelligent ApplicatioN-Oriented System
IAS	Institute for Advanced Study, Princeton, NJ
Illiac IV	First array computer
Infiniband	High speed internode communication network
IT	Information Technology

J	Energy consumption of a node
j	Counter for configuration
JSDL	Job Submission Description Language

k	Counter for application component
K_d	Data transfer costs
K_e	CPU time costs
K_{eco}	Energy costs
K_ℓ	License fees
K_w	Costs due to waiting time

| ℓ | Counter for the Grid sites |
| L_i | Latency time of the network |

ℓ_i	Number of links in the network
L_m	Latency time of the main memory system
LAN	Local Area Network
LARC	Livermore Atomic Research Computer
LEMan	Linear ElectroMagnetic waves code for plasma heating
Linpack	HPL benchmark for the TOP500 list
Linux	Openware UNIX based operating system
LISP	High-level programming language
LSI	Large System Integration
M	Number of resources in a site
M_i	Total memory size of a resource
M_m	Sustained main memory bandwidth of a node
m_{node}	Memory size attached to a node
M_∞	Peak memory bandwidth of a node
MADAM	Manchester electronic computer
MARK	Harvard computer
MB/s	10^6 Bytes/second
Memcom	Data management system by SMR
MF/s	10^6 floating point operations per second
MHD	MagnetoHydroDynamics
MIPJ	Million Instructions per Joule
MM	Monitoring Module
Moore's law	Quadruples every 3 years
MPI	Message Passing Interface
MPP	Massively Parallel Processors
ms	milliseconds
MS	Monitoring Service
MSS	Meta-Scheduling Service
MW	10^6 Watt
Mw/s	10^6 64bit words per second
Myrinet	High speed internode communication network
N	Total number of elements $= N_x * N_y * N_z$
N_{conf}	Number of eligible configurations
N_{core}	Number of cores in a processor
N_{CPU}	Number of processors in a node
$N_{diff,k}$	Number of mesh points contributing to the difference scheme
$N_{node,i}$	Number of nodes in a resource
N_s	Number of connectivity surfaces
n_{sd}	Number of subdomains
N_{site}	Number of sites in a Grid
N_{var}	Number of variables per mesh point. Also number of PDEs
N_x	Number of elements in x-direction
N_y	Number of elements in y-direction

N_z	Number of elements in z-direction
NEC SX	Vector computer series by Nippon Electric Company
NIC	Network Interface Connection
node	Computational unit connected to a network
NORC	Naval Ordnance Research Calculator
NOW	Network of Workstations
NTDS	US Navy special purpose computer (1957) built by Univac
NUMA	Non-Uniform Memory Access
O	Total number of operations performed in one node
Occam	Parallel library for Transputers
OpenMP	Directives to parallelize on an SMP or NUMA node
Opteron	Multi-processor, multi-core node by AMD
p	Polynomial degree
P	Gas pressure in SpecuLOOS
P_i	Total number of *nodes* of resource r_i
$P_{k,i}$	Number of *nodes* reserved by a_k in resource r_i
$P_{\frac{1}{2}}$	Number of nodes giving half of ideal speedup
PAPI	Performance Application Programming Interface
PC	Personal computer
PCI	Peripheral Component Interconnect
PDE	Partial Differential Equation
Pentium4	Intel node with $N_{CPU}=N_{core}=1$
Perfmon	Performance monitor
PF/s	10^{15} floating point operations per second
Plankalkül	First arithmetic language
PMPI	Profiled MPI
PPE	Power Processor Element in a CELL
PS3	Playstation 3 by Sony
PVM	Parallel Virtual Machine
P2P	Pear to Pear
QoS	Quality of Service
r_a	Nodal peak performance of an application
R_a	Effective nodal performance of an application
r_f	Bus-to-core ratio (frequency)
$(r_f; r_v)$	P-State
$(r_{f,c}; r_{v,c})$	C-State
r_i	Resource number i in a Grid
R_i	Total peak performance of a resource
$r_{k,i}$	Effective performance of an application
r_v	Bus-to-core ratio (voltage)

R_∞	Peak performance of a *node*
RAM	Random Access Memory
RB	Resource Broker
Resource	A node, SMP, NUMA, cluster, MPP
RISC	Reduced Instruction Set Computer
RMS	Resource Management System
RRDB	Round Robin Data Base
S_a^i	Management overhead
S_b^i	Interests of investment
S_c^i	Investment costs
S_g^i	Margin
S_j^i	Insurance costs
S	Total number of words sent by the component
σ	Average message size S/Z
S_l^i	Infrastructure costs
s_ℓ	Site in a Grid
S_m^i	Maintenance fees
S_p^i	Personnel costs
SAXPY	Blas1 operation
ScalaPACK	Scalable linear algebra library package
SDRAM	Synchronous Dynamic Random Access Memory
SI	System Information
SIMD	Single Instruction Multiple Data
Site	Locally administrated set of resources
SMP	Symmetric Multi Processor
SMXV	Sparse matrix times vector multiply benchmark
SPE	Synergistic Processing Element in a CELL
SpecuLOOS	Spectral element code in CFD
SSE	Streaming SIMD Extensions
SwiNG	Swiss National Grid initiative
T	Total execution time
t^0	Submission time of a_k
T_1	Total computing time of one node
t_b	Data transfer time
t_{comm}	Internode communication time
t_{comp}	Computation time
$t_{k,i}^d$	Time after having saved data
$t_{k,i}^e$	End time of execution for a_k using $P_{k,i}$ nodes on resource r_i
t_L	Latency time in μs
$t_{k,i}^r$	Time at which the job disappears from the system
$t_{k,i}^s$	Start time of execution for a_k using $P_{k,i}$ nodes on resource r_i
TMC	Connection Machines Corporation

T_P	Computing time on P nodes
TB	10^{12} Bytes
TCP/IP	Standard network protocol
TF/s	10^{12} floating point operations per second
TNet	Internode communication network by Supercomputing Systems
TOP500	Performance list based on Linpack benchmark
ULSI	Ultra Large System Integration
UNICORE	Grid middleware
V_a	O/W
V_c	R_∞/b_i
V_{cc}	Processor core voltage
V_m	$g_m V_\infty$
V_∞	R_∞/M_∞
VAMOS	Veritable Application MOnitoring System
VHSIC	Very High System Integration Circuitry
VLSI	Very Large System Integration
ω_{core}	Number of functional units in a core
W	Number of main memory accesses in a node
WAN	Wide Area Network
Windows	Operating system by Microsoft
Woodcrest	Intel Xeon node with $N_{CPU}=2$, N $N_{core}=2$ or 4
W	Watt
WS	Web Service
WSAG	WS AGreement
WSRF	WS Resource Framework
Z	Cost function value
Z	Number of messages sent over the network
Z4	Zuse machine at ETHZ

References

1. AHUSBORDE, E., GRUBER, R., AZAÏEZ, M., AND SAWLEY, M. L. Physics-conforming constraints-oriented numerical method. *Physical Review E 75*, 5 (2007), 056704.
2. AIELLO, M., FRANKOVA, G., AND MALFATTI, D. What's in an agreement? an analysis and an extension of ws-agreement. In *Service-Oriented Computing - ICSOC 2005*. Springer Berlin / Heidelberg, 2005, pp. 424–436.
3. ALDER, B., AND WAINWRIGHT, T. E. Studies in molecular dynamics. 1. general method. *J. Chem. Phys. 31*, 2 (1959), 459–466.
4. AMD CORP. AMD Cool'n'Quiet 2.0. Technology Overview, 2008.
5. ANDERSON, D., COOPER, W., GRUBER, R., MERAZZI, S., AND SCHWENN, U. Methods for the efficient calculation of the MHD stability properties of magnetically confined fusion plasmas. *Int. J. of Supercomputer Applications 4*, 3 (1990), 34–47.
6. ANDERSON, D., COOPER, W. A., GRUBER, R., AND SCHWENN, U. Gigaflop Performance Award. Cray Research Inc., 1989.
7. ANDERSON, E., BAI, Z., BISCHOF, C., BLACKFORD, L., DEMMEL, J., DONGARRA, J., DU CROZ, J., HAMMARLING, S., GREENBAUM, A., MCKENNEY, A., AND SORENSEN, D. *LAPACK Users' guide*, 3td ed. Society for Industrial and Applied Mathematics, Philadelphia, PA, USA, 1999.
8. ANDREOZZI, S., BURKE, S., FIELD, L., FISHER, S., K'ONYA, B., MAMBELLI, M., SCHOPF, J., VILJOEN, M., AND WILSON, A. Glue schema specification version 1.2, 2008.
9. ANJOMSHOAA, A., BRISARD, F., DRESCHER, M., FELLOWS, D., LY, A., MCGOUGH, A., PULSIPHER, D., AND SAVVA, A. Job submission description language (JSDL) specification, version 1.0., 2008.
10. APPERT, K., BERGER, D., GRUBER, R., AND RAPPAZ, J. A new finite element approach to the normal mode analysis in MHD. *J. Comp. Phys. 18* (1975), 284–299.
11. BABBAGE, C. A note respecting the application of machinery to the calculation of astronomical tables. *Mem. Astron. Soc. 1* (1822), 309.
12. BARNES, G., BROWN, R., KATO, M., KUCK, D., SLOTNICK, D., AND STOKES, R. The *ILLIAC IV* computer. *IEEE Trans. Comput. C-29* (1968), 746–757.
13. BARNES, J. *ADA 95*, 2nd ed. Addison-Wesley, 1998.
14. BATCHER, K. The *STARAN* computer. *Supercomputers 2* (1979), 33–49.
15. BELL, N., AND GARLAND, M. Efficient sparse matrix-vector multiplication on CUDA. *NVIDIA Technical Report NVR-2008-004* (2008).
16. BIANCHINI, R., AND RAJAMONY, R. Power and energy management for server systems, 2003.
17. BLACKBURN, M. Five ways to reduce data center server power consumption. Tech. Rep. Whitepaper Number 7, the green grid, 2008.
18. BLOCH, J. *Effective Java programming language guide*. Addison-Wesley, 2001.
19. BOESCH, F., AND TINDELL, R. Circulants and their connectivities. *J. of Graph Theory 8* (1984), 487–499.

20. BONOMI, E., FLÜCK, M., GEORGE, P., GRUBER, R., HERBIN, R., PERRONNET, A., MERAZZI, S., RAPPAZ, J., RICHNER, T., SCHMID, V., STEHLIN, P., TRAN, C., VIDRASCU, M., VOIROL, W., AND VOS, J. ASTRID: Structured finite element and finite volume programs adapted to parallel vectorcomputers. *Computer Physics Reports 11* (1989), 81–116.

21. BOOZER, A. Guiding center drift equations. *Phys. Fluids 23* (1980), 904.

22. BOSSHARD, C., BOUFFANAIS, R., CLÉMENÇON, C., DEVILLE, M., FIÉTIER, N., GRUBER, R., KEHTARI, S., KELLER, V., AND LÄTT, J. Computational performance of a parallelized three-dimensional high-order spectral element toolbox. *Proceedings of the APPT09 conference, Rapperswil* (2009).

23. BOUFFANAIS, R., AND DEVILLE, M. Mesh update techniques for free-surface flow solvers using spectral element method. *J. Sci. Comput. 27* (2006), 137–149.

24. BOUFFANAIS, R., DEVILLE, M., FISCHER, P., LERICHE, E., AND WEILL, D. Large-eddy simulation of the lid-driven cubic cavity flow by the spectral element method. *J. Sci. Comput. 27* (2006), 151–162.

25. BOUFFANAIS, R., DEVILLE, M., AND LERICHE, E. Large-eddy simulation of the flow in a lid-driven cubical cavity. *Phys. Fluids 19* (2007). Art. 055108.

26. BRANDT, A. Multi-level adaptive techniques (*MLAT*) for fast numerical solutions to boundary value problems. In *Lecture Notes in Physics 18* (1973), H. Cabannes and R. Temam, Eds.

27. BRODOWSKI, D. cpufrequtils. CPU frequence management tool.

28. BROMLEY, A. Grand challenges 1993: High performance computing and communications. a report by the committee on physical, mathematical, and engineering sciences. the fy 1993 u.s. research and development program. Tech. rep., National Science Foundation, 1992.

29. BUTTARI, A., LUSZCZEK, P., KURZAK, J., DONGARRA, J., AND G., B. SCOP3: a rough guide to scientific computing on the PlayStation 3. Tech. rep., Innovative Computing Laboratory, University of Tennessee Knoxville, 2007. UT-CS-07-595.

30. CALINGER, R. *A Contextual History of Mathematics*. Addison-Wesley, 1999.

31. CAMPBELL, M., HOANE, JR., A. J., AND HSU, F.-H. Deep blue. *Artif. Intell. 134*, 1-2 (2002), 57–83.

32. CANUTO, C., HUSSAINI, M., QUARTERONI, A., AND ZANG, T. *Spectral Methods in Fluid Dynamics*. Springer, New York, 1988.

33. CAR, R., AND PARRINELLO, M. Unified approach for molecular dynamics and density-functional theory. *Phys. Rev. Lett. 55* (1985), 2471–2474.

34. COHEN, D. On holy wars and a plea for peace. *Computer 14*, 10 (1981), 48–54.

35. COOLEY, J., AND TUKEY, J. An algorithm for the machine calculation of complex Fourier series. *Mathematics of Computation 19*, 90 (1965), 297–301.

36. COURANT, R. Variational methods for the solution of problems of equilibrium and vibration. *Bull. of American Math. Society 49* (1943), 1–61.

37. COUZY, W., AND DEVILLE, M. Spectral-element preconditioners for the Uzawa pressure operator applied to incompressible flows. *J. Sci. Comput. 9* (1994), 107–112.

38. CRISTIANO, K., GRUBER, R., KELLER, V., KUONEN, P., MAFFIOLETTI, S., NELLARI, N., SAWLEY, M.-C., SPADA, M., TRAN, T.-M., WÄLDRICH, O., WIEDER, P., AND ZIEGLER, W. Integration of ISS into the VIOLA Meta-scheduling Environment. In *Proceedings of the Integrated Research in Grid Computing Workshop* (November 2005), S. Gorlatch and M. Danelutto, Eds., Universit di Pisa, pp. 357–366.

39. CRISTIANO, K., KUONEN, P., GRUBER, R., KELLER, V., SPADA, M., TRAN, T.-M., MAFFIOLETTI, S., NELLARI, N., SAWLEY, M.-C., WÄLDRICH, O., ZIEGLER, W., AND WIEDER, P. Integration of ISS into the VIOLA Meta-Scheduling Environment. Tech. Rep. TR-0025, Institute on Resource Management and Scheduling, CoreGRID - Network of Excellence, January 2006.

40. CULLER, D., SINGH, J., AND GUPTA, A. *Parallel computer architecture*. Morgan Kaufmann Publishers, 1999.

41. DANTZIG, G. *Linear programming and extensions*. Princeton University Press, 1963.
42. DEGTYAREV, L., MARTYNOV, A., MEDVEDEV, S., TROYON, F., VILLARD, L., AND GRUBER, R. The KINX ideal MHD stability code for axisymmetric plasmas with separatrix. *Comp. Phys. Comm. 103* (1997), 10–24.
43. DEVILLE, M., FISCHER, P., AND MUND, E. *High-Order Methods for Incompressible Fluid Flow*. Cambridge Monographs on Applied and Computational Mathematics. Cambridge University Press, 2002.
44. DROTZ, A., GRUBER, R., KELLER, V., THIÉMARD, M., TOLOU, A., TRAN, T.-M., CRISTIANO, K., KUONEN, P., WIEDER, P., WÄLDRICH, O., ZIEGLER, W., MANNEBACK, P., SCHWIEGELSHOHN, U., YAHYAPOUR, R., KUNSZT, P., MAFFIOLETTI, S., SAWLEY, M.-C., AND WITZIG, C. Application-oriented scheduling for HPC Grids. Tech. Rep. TR-0070, Institute on Resource Management and Scheduling, CoreGRID - Network of Excellence, February 2007.
45. DUBOIS-PÉLERIN, Y. SpecuLOOS: An object-oriented toolbox for the numerical simulation of partial differential equations by spectral and mortar element method. Tech. rep., LMF,Swiss Federal Institute of Technology, Lausanne, 1998.
46. DUBOIS-PÉLERIN, Y., VAN KEMENADE, V., AND DEVILLE, M. An object-oriented toolbox for spectral element analysis. *J. Sci. Comput. 14*, 1 (1999), 1–29.
47. EGLI, W., KOGLSCHATZ, U., GERTEISEN, E., AND GRUBER, R. 3D computation of corona, ion induced secondary flows and particle motion in technical ESP configurations. *J. of electrostatics 40,41* (1997), 425–430.
48. FENG, W., AND CAMERON, K. The GREEN500 List: Encouraging sustainable supercomputing. *Computer 40*, 12 (2007), 50–55.
49. FISCHER, P. An overlapping Schwarz method for spectral element solution of the incompressible Navier-Stokes equations. *J. Comp. Phys. 133* (1997), 84–101.
50. FISCHER, P., AND PATERA, A. Parallel spectral element solution of the Stokes problem. *J. Comput. Phys. 92*, 2 (1991), 380–421.
51. FLINN, J., AND SATYANARAYANAN, M. Energy-aware adaptation for mobile applications. In *SOSP '99: Proceedings of the seventeenth ACM symposium on Operating systems principles* (New York, NY, USA, 1999), ACM, pp. 48–63.
52. FOSTER, I., AND KESSELMAN, C. *The Grid: Blueprint for a New Computing Infrastructure*. Morgan Kaufmann Publishers Inc., San Francisco, CA, USA, nov 1998.
53. FOSTER, I., AND KESSELMAN, C. *The Grid 2: Blueprint for a New Computing Infrastructure*. Morgan Kaufmann Publishers Inc., San Francisco, CA, USA, 2003.
54. FUJIMOTO, N. Faster matrix-vector multiplication on GeForce 8800GTX. *The proceedings of IEEE International Parallel and Distributed Processing (IPDPS)* (2008).
55. GARDNER, M. The Abacus: primitive but effective digital computer. *Scientific American 222*, 1 (1970), 124–126.
56. GARDNER, M. The calculating rods of John Napier, the eccentric father of the logarithm. *Scientific American 228*, 3 (1973), 110–113.
57. GOEDECKER, S., AND HOISIE, A. *Performance optimisation of numerically intensive codes*. SIAM, 2001.
58. GOLUB, G., AND VAN LOAN, C. *Matrix Computations*, 2nd ed. John Hopkins University Press, London, 1989.
59. GOTTLIEB, D., AND ORSZAG, S. *Numerical Analysis of Spectral Methods : Theory and Applications*. SIAM-CBMS, Philadeliphia, 1977.
60. GROPP, W., LUSK, E., AND SKJELLUM, A. *Using MPI: Portable parallel programming with the message-passing interface*, 2nd ed. MIT Press, Cambridge, MA, USA, 1999.
61. GRUBER, R. Finite hybrid elements to compute the ideal MHD spectrum of an axisymmetric plasma. *J. Comp. Phys. 26* (1978), 379–389.
62. GRUBER, R., COOPER, W., BENISTON, M., GENGLER, M., AND MERAZZI, S. Software development strategies for parallel computer architectures. *Physics Reports 207* (1991), 167–214.

63. GRUBER, R., KELLER, V., LERICHE, E., AND HABISREUTINGER, M.-A. Can a Helmholtz solver run on a cluster? *Cluster Computing, 2006 IEEE International Conference on* (25-28 Sept. 2006), 1–8.

64. GRUBER, R., KELLER, V., MANNEBACK, P., THIÉMARD, M., WALDRICH, O., WIEDER, P., AND ZIEGLER, W. Integration of Grid Cost Model into ISS/VIOLA Meta-Scheduler Environment. In *Proceedings of the Euro-Par 2006* (Dresden, Germany, August 2007), LNCS, Springer.

65. GRUBER, R., MERAZZI, S., COOPER, W., FU, G., SCHWENN, U., AND ANDERSON, D. Ideal MHD stability computations for 3D magnetic fusion devices. *Computer Methods in Applied Mechanics and Engineering 91* (1991), 1135–1149.

66. GRUBER, R., AND RAPPAZ, J. *Finite Element Methods in Linear Ideal MHD.* Springer Series in Computational Physics. Springer, Berlin, 1985.

67. GRUBER, R., AND TRAN, T.-M. Scalability aspects on commodity clusters. *EPFL Supercomputing Review 14* (2004), 12–17.

68. GRUBER, R., TROYON, F., BERGER, D., BERNARD, L., ROUSSET, S., SCHREIBER, R., KERNER, W., SCHNEIDER, W., AND ROBERTS, K. ERATO stability code. *Computer Physics Communications* (1981), 323–387.

69. GRUBER, R., VOLGERS, P., DE VITA, A., STENGEL, M., AND TRAN, T.-M. Parameterisation to tailor commodity clusters to applications. *Future Generation Computer Systems 19* (2003), 111–120.

70. HACKBUSCH, W. *Multigrid Methods and Applications.* Springer Series in Computational Mathematics 4. Springer, New York, 1985.

71. HALDENWANG, P., LABROSSE, G., ABBOUDI, S., AND DEVILLE, M. Chebyshev 3D spectral and 2D pseudospectral solvers for the Helmholtz equation. *J. Comput. Phys. 55* (1984), 115–128.

72. HARTREE, D. The *ENIAC*, an electronic computing machine. *Nature 158* (1946), 500–506.

73. HESTENES, M., AND STIEFEL, E. Methods of conjugate gradients for solving linear systems. *J. Res. Nat.Bur. Stand. 49* (1952), 409–436.

74. HILLIS, W. D. *The connection machine.* PhD thesis, MIT, 1985.

75. HILLIS, W. D., AND TUCKER, L. W. The CM-5 connection machine: a scalable supercomputer. *Commun. ACM 36*, 11 (1993), 31–40.

76. HIRSHMAN, S. P., SCHWENN, U., AND NÜHRENBERG, J. Improved radial differencing for 3D MHD calculations. *J. Comp. Phys. 87* (1990), 396–407.

77. HOCKNEY, R. The potential calculation and some applications. *Methods of Comput. Phys. 9* (1970), 135–211.

78. HOCKNEY, R., AND JESSHOPE, C. *Parallel Computers 2*, 2nd ed. Adam Hilger, Bristol, 1988.

79. HOUSEHOLDER, A. Unitary triangularization of a nonsymmetric matrix. *ACM 5* (1958), 339–342.

80. INTEL CORP. *Intel 64 and IA-32 Architectures Software Developer's Manuals. Volumes 1, 2A, 2B, 3A, 3B, see http://software.intel.com/en-us/intel-compilers.*

81. INTEL CORP. Enhanced Intel speedstep technology for the Pentium M processor. Tech. Rep. Whitepaper, Intel, March 2004.

82. INTERNATIONAL ORGANIZATION FOR STANDARDIZATION. *Information technology—open systems interconnection—basic reference model: The basic model. Standard ISO/IEC 7498-1*, November 1994.

83. INTERNATIONAL ORGANIZATION FOR STANDARDIZATION. *Information Technology: Programming Languages: Fortran. Part 1: Base Language, ISO/IEC 1539-1:1997(E)*, 15 December 1997.

84. ITER. International Torus Experimental Reactor, http://www.iter.org.

85. IVERSON, K. *A programming language.* Wiley, London, 1962.

86. JACOBI, C. Über ein leichtes Verfahren, die in der Theorie der Säcularstörungen vorkommenden Gleichungen numerisch aufzulösen. *Crelle's J. 30* (1846), 51–94.

87. KATZ, V. *A History of Mathematics*. Addison-Wesley, 2004.

88. KELLER, V. *Optimal application-oriented resource brokering in a high performance computing Grid*. PhD thesis, École Polytechnique Fédérale de Lausanne, 2008. Number 4221.

89. KELLER, V., PURAGLIESI, R., JUCKER, M., AND PENA, G. Optimisation et parallélisme avec OpenMP : l'approche grain fin. *Flash Informatique EPFL FI 3/07* (Mar 2007).

90. KERNIGHAN, B., AND RITCHIE, D. *The C Programming Language*. Prentice-Hall, 1988.

91. KUONEN, P. The K-ring: a versatile model for the design of MIMD computer topology. In *Proc. of High performance Computing Conference (HPC'99)* (1995), pp. 381–385.

92. LANCZOS, C. An iterative method for the solution of the eigenvalue problem of linear differential and integral operators. *J. Res. Nat. Bur. Stand. 45*, 4 (1950), 255–282.

93. LAWSON, C., HANSON, R., KINCAID, D., AND KROGH, F. Basic Linear Algebra Subprograms for Fortran usage. *ACM Transactions on Mathematical Software 5*, 3 (September 1979), 308–323.

94. LEISERSON, C. Fat-Tree, Universal network for hardware-efficient supercomputing. *IEEE Transactions on Computers 34*, 10 (1985), 892–901.

95. LERICHE, E. Direct numerical simulation in a lid-driven cubical cavity at high Reynolds number by a Chebyshev spectral method. *J. Sci. Comput. 27*, 1-3 (2006), 335–345.

96. LERICHE, E., AND GAVRILAKIS, S. Direct numerical simulation of the flow in a lid-driven cubical cavity. *Phys. Fluids 12* (June 2000), 1363–1376.

97. LERICHE, E., PERCHAT, E., LABROSSE, G., AND DEVILLE, M. O. Numerical evaluation of the accuracy and stability properties of high-order direct Stokes solvers with or without temporal splitting. *J. Sci. Comput. 26*, 1 (2006), 25–43.

98. LINDHOLM, E., NICKOLLS, J., OBERMAN, S., AND J., M. NVIDIA Tesla: A unified graphics and computing architecture. *IEEE Micro, 28*, 2 (2008), 39–55.

99. LOURAKIS, M. levmar: Levenberg-Marquardt nonlinear least squares algorithms in C/C++. website, Jul. 2004.

100. LYNCH, R., RICE, J., AND THOMAS, D. Direct solution of partial difference equations by tensor product methods. *Numerische Mathematik 6* (1964), 185–199.

101. MADSEN, K., NIELSEN, H., AND TINGLEFF, O. *Methods for Non-Linear Least Squares Problems*, 2nd ed. Informatics and Mathematical Modelling, Technical University of Denmark, DTU, Richard Petersens Plads, Building 321, DK-2800 Kgs. Lyngby, 2004.

102. MAY, D., AND TAYLOR, R. Occam - an overview. *Microprocessors and microsystems 8* (1984), 73–79.

103. MERAZZI, S. *MEMCOM User Manual*, 1998. SMR SA, Bienne, Switzerland.

104. METCALF, M., AND REID, J. *Fortran 90/95 explained*, 2nd ed. Oxford University Press, 1999.

105. METROPOLIS, N., HOWLETT, J., AND ROTA, G.-C. *A history of computing in the 20th century*. Academic Press, 1976.

106. METROPOLIS, N., ROSENBLUTH, A., ROSENBLUTH, M., TELLER, A., AND TELLER, E. Equation of state calculations by fast computing machines. *J. of Chem. Phys. 21*, 6 (1953), 1087–1092.

107. MUDGE, T. Power: A first class design constraint for future architecture and automation. In *HiPC '00: Proceedings of the 7th International Conference on High Performance Computing* (London, UK, 2000), Springer-Verlag, pp. 215–224.

108. NAUR ET AL., P. Report on the algorithmic language Algol60. *Commun. of the ACM 3*, 5 (1960), 299–314.

109. POPOVICH, P., COOPER, W., AND VILLARD, L. A full-wave solver of the Maxwell's equations in 3D cold plasmas. *Comput. Phys. Comm.* (2006), 175–250.

110. RASHEED, H., GRUBER, R., KELLER, V., KUONEN, P., WALDRICH, O., WIEDER, P., AND ZIEGLER, W. IANOS: An Intelligent Application Oriented Scheduling Framework For An HPCN Grid. In *Grid Computing* (2008), pp. 237–248.

111. RITCHIE, D., AND THOMPSON, K. The *UNIX* time-sharing system. *C. ACM 17*, 1 (1974), 365–437.

112. RUTISHAUSER, H. Solution of eigenvalue problems with the LR transformation. *Appl. Math. Series 49* (1958), 47–81.

113. RUTISHAUSER, H. *Lectures on Numerical Mathematics*. Birkhäuser, Basel, 1990.

114. SCOTT, S. Synchronization and communication in the T3E multiprocessor. In *Proceedings of the 7th International Conference on architectural support for programming languages and operating systems* (1996), vol. 1, pp. 26–36.

115. SHARMA, S., HSU, C.-H., AND FENG, W.-C. Making a case for a GREEN500 List. *Parallel and Distributed Processing Symposium, 2006. IPDPS 2006. 20th International* (April 2006), 8 pp.–.

116. SHIMPI, A., AND WILSON, D. NVIDIA's 1.4 Billion Transistor GPU: GT200 Arrives as the GeForce GTX 280 and 260.

117. SNOWDON, D., RUOCCO, S., AND HEISER, G. Power management and dynamic voltage scaling: Myths and facts. In *Proceedings of the 2005 Workshop on Power Aware Real-time Computing* (Sep 2005).

118. SON, S., MALKOWSKI, K., CHEN, G., KANDEMIR, M., AND RAGHAVAN, P. Reducing energy consumption of parallel sparse matrix applications through integrated link/CPU voltage scaling. *J. Supercomput. 41*, 3 (2007), 179–213.

119. STERLING, T., SALMON, J., BECKER, D. J., AND SAVARESE, D. F. *How to build a Beowulf*. The MIT Press, 1999.

120. STERN, N., STERN, R., AND LEY, J. *Cobol for the 21st century*, 10th ed. John Wiley and Sons, 2002.

121. STEWARD, I. Gauss. *Scientific American 237*, 1 (1977), 122–131.

122. STROUSTRUP, B. *The C++ Programming Language*. Addison-Wesley, 1997.

123. STRZODKA, R. WS 2008, lecture and course: Massively Parallel Computing with CUDA, 2008.

124. SUNDARAM, V. *PVM: A framework for parallel computing. Concurrency: Practice and Experience 2*, 4 (1990), 313–339.

125. SUSE. SuSE LINUX. OpenSUSE, 2003.

126. TANENBAUM, A. *Structured Computer Organisation*, fourth ed. Prentice Hall, 1999.

127. TAUBES, G. The rise and fall of thinking machines. *Inc. The daily resource for entrepreneurs* (1995).

128. THE EUROPEAN COMMISSION. The European Economic and Monetary Union (EMU).

129. THE STAFF OF THE COMPUTATION LABORATORY. A manual of operation for the automatic sequence controlled calculator. Tech. rep., Harvard University, 1946.

130. TRAN, T.-M., WHALEY, D., MERAZZI, S., AND GRUBER, R. DAPHNE, a 2D axisymmetric electron gun simulation code. *Proc.6th joint EPS-APS Int. Conf. on Physics Computing* (1994), 491–494.

131. TROYON, F., GRUBER, R., SAURENMANN, H., SEMENZATO, S., AND SUCCI, S. MHD limits to plasma confinement. *Plasma Phys. 26*, 1A (1983), 209–215.

132. TURING, A. M. Computing machinery and intelligence. *Mind LIX* (1950), 433–460.

133. VAN KEMENADE, V. Incompressible fluid flow simulation by the spectral element method. Tech. Rep. FN 21-40'512.94 project, IMHEF–DGM, Swiss Federal Institute of Technology, Lausanne, 1996.

134. VLADIMIROVA, M., STENGEL, M., TRIMARCHI, G., MERIANI, S., BARTH, J., SCHNEIDER, W.-D., DE VITA, A., AND BALDERESCHI, A. Using massively parallel computing to model molecular adsorption and self-assembly on noble metal surfaces. *EPFL Supercomputing Review 11* (2002), 23–32.

135. VON NEUMANN, J. First draft of a report on the EDVAC. *IEEE Ann. Hist. Comput. 15*, 4 (1993 (1st publication 1945)), 27–75.

136. WÄLDRICH, O., WIEDER, P., AND ZIEGLER, W. A meta-scheduling service for co-allocating arbitrary types of resources. *Lecture Notes in Computer Science 3911* (2006), 782–791.

137. WEISER, M., WELCH, B., DEMERS, A., AND SHENKER, S. Scheduling for reduced CPU energy. In *OSDI '94: Proceedings of the 1st USENIX conference on Operating Systems Design and Implementation* (Berkeley, CA, USA, 1994), USENIX Association, p. 2.

138. WILKES, M. V. The *EDSAC*, an electronic calculating machine. *J. Sci. Instr. 26* (1949), 385.

139. WILKINSON, J. H. Error analysis of direct methods of matrix inversion. *J. ACM 8*, 3 (1961), 281–330.

140. WILLIAMS, F. C. Cathode ray tube storage. Report on a Conference on High Speed Automatic Computing Machines, 1949.

141. WILLIAMS, F. C. *MADAM*, automatic digital computation. In *Proceedings of a Symposium held at National Physical Laboratory, United Kingdom* (1953).

142. WILLIAMS, M. *History of Computing Technology*, 2nd ed. IEEE Computer Society, 1997.

143. WIRTH, N. *Programming in Modula-2*. Springer Verlag, 1985.

144. WIRTH, N., AND JENSEN, K. *PASCAL-User manual and report*. Springer Verlag, 1974.

About the authors

Ralf Gruber

Ralf Gruber is a physicist born in 1943 in Zurich, Switzerland. After physics studies at the ETH Zurich he worked for Rieter AG in Winterthur (Switzerland) on optimization problems related to spinning machines. He received his PhD degree at École Polytechnique Fédérale de Lausanne (EPFL) in 1976 in the plasma physics field. Between 1985 and 1990 he created and leads the GASOV-EPFL, a user support group for the Cray-1 and the Cray-2 machines that were just installed at EPFL, and was responsible for the ERCOFTAC coordination center. His strong relation to the applications and the users did not stop. Between 1991 and 1995 he was vice-president of the Swiss National Supercomputing Centre (CSCS) in Manno (Switzerland) where he creates the NEC SX-3 user support group, and built up co-operations with industrial partners. In 1997 he was director of the Swiss-TX project : a high performance computing machine based on commodity nodes and high performance interconnection network (T-Net). Together with Bill Camp he founds the yearly SOS workshops in 1997. From 1996 to 2002, he leads the CAPA (for "Centre pour les Applications Parallèles et Avancées" or Parallel and Advanced Applications Center) at EPFL. He gives a postgrade lecture on "High Performance Computing Methods" from where parts of that book is largely inspired. From 2005, he turns to Grid computing with the launching of the Intelligent Scheduling System (ISS) which became in 2008 the ïanos framework.

Vincent Keller

Vincent Keller is a computer scientist born in 1975 in Lausanne, Switzerland. He received his Master degree in Computer Science from the University of Geneva (Switzerland) in 2004. From 2004 to 2005, he holds a full-time researcher position at the University Hospital of Geneva (HUG). He was involved at HUG on simulating blood flows in cerebral aneurysms using real geometries constructed from 3D X-rays tomography slices. The numerical method used was Lattice-Boltzman Method (LBM). He received his PhD degree in 2008 from the École Polytechnique

Fédérale de Lausanne (EPFL) in the HPCN and HPC Grids fields. His supervisor was Dr. Ralf Gruber. He developed a low-level application-oriented monitoring system (VAMOS) and the Resource Broker of the ïanos (Intelligent ApplicatioN-Oriented System) framework. Since 2009, he holds a full-time researcher position at Fraunhofer SCAI (University of Bonn). His research interests are in HPC applications analysis, Grid and cluster computing and energy efficiency of large computing ecosystems.

Index